Ginger - Cultivation and Use

Edited by Prashant Kaushik and Rabia Shabir Ahmad

Published in London, United Kingdom

Ginger - Cultivation and Use
http://dx.doi.org/10.5772/intechopen.102135
Edited by Prashant Kaushik and Rabia Shabir Ahmad

Contributors
Latika Yadav, Upasana, Zukile Mbita, Lilian Makgoo, Mohini Chandrashekhar Upadhye, Uday Deokate, Rabia Shabir Ahmad, Muhammad Imran, Muhammad Kamran Khan, Muhammad Haseeb Ahmad, Ali Imran, Huda Ateeq, Madonna Ngwatshipane Mashabela, Wilfred Otang-Mbeng, Douglas Ongeri Ochora, Temitayo Esther AdeyeOluwa, N. Divyashree, S. Poojashree, S. Venukumar, Y.C. Vishwanath, Rakesh Chandra Nainwal, Shri Krishna Tewari, Saja Shareef, Prashant Kaushik, Navjot Singh Brar, Prakash Mahala, Kartik Sharma, Parmdeep Singh Dhanda, Alpa Yadav, Meenakshi Sharma

First published in London, United Kingdom, 2023 by IntechOpen
IntechOpen is the global imprint of INTECHOPEN LIMITED, registered in England and Wales, registration number: 11086078, 5 Princes Gate Court, London, SW7 2QJ, United Kingdom

British Library Cataloguing-in-Publication Data
A catalogue record for this book is available from the British Library

Additional hard and PDF copies can be obtained from orders@intechopen.com

Ginger - Cultivation and Use
Edited by Prashant Kaushik and Rabia Shabir Ahmad
p. cm.
Print ISBN 978-1-83969-118-8
Online ISBN 978-1-83969-119-5
eBook (PDF) ISBN 978-1-83969-157-7

Meet the editors

Dr. Prashant Kaushik holds Adjunct Faculty positions at three universities and is also a genome data engineer in a private company. He holds a BS (Hons) in agriculture, a PG diploma in floriculture and landscaping, an MS in vegetable science, and an MBA in production and operations management. After obtaining a Ph.D. in biotechnology from Universitat Politecnica de Valencia, Spain, Dr. Kaushik was a bioinformatician at Nagano University in Ueda, Japan and worked at Yokohama Ueki in Kikugawa City, Japan. Dr. Kaushik has published 110 SCI or SCI-E Q1, Q2, and Q3 research articles, and an equivalent number of non-indexed research publications. He has written six books and 14 book chapters. Among his research interests are molecular biology, plant breeding, genomics, and population genetics.

Dr. Rabia Shabir Ahmad is an associate professor with a strong background in academics, teaching, and research. She successfully completed doctoral research funded by the Indigenous Fellowship Program, Higher Education Commission (HEC), Pakistan. During her academic career, Dr. Ahmad was awarded and successfully completed a Start-Up Research Grant Program (SRGP) and National Research Program for Universities (NRPU) project from the HEC as Principal Investigator on honey and its analysis through high-performance liquid chromatography (HPLC). Along with her teaching and research supervising responsibilities, Dr. Ahmad is also a reviewer for different research journals. She has published numerous research papers in international and national journals and edited several books.

Contents

Preface **IX**

Section 1
Medicinal Properties **1**

Chapter 1 **3**
Introductory Chapter: Curcumin and Its Therapeutic Potency
by Rabia Shabir Ahmad, Muhammad Imran, Muhammad Kamran Khan, Muhammad Haseeb Ahmad, Ali Imran and Huda Ateeq

Chapter 2 **9**
Anti-Inflammatory and Antioxidant Activities of Ginger
by Saja Shareef

Chapter 3 **17**
Antioxidant Potential of Phytoconstituents with Special Emphasis on Curcumin
by Uday Deokate and Mohini Upadhye

Chapter 4 **33**
Pharmacological Properties of Ginger Combinations
by Douglas Ongeri Ochora

Chapter 5 **47**
Golden Spice Turmeric and Its Health Benefits
by Latika Yadav and Upasana

Chapter 6 **63**
The Therapeutic and Phytopharmacological Potential of Ginger (*Zingiber officinale*)
by Madonna Ngwatshipane Mashabela and Wilfred Otang-Mbeng

Chapter 7 **79**
Germicidal and Antineoplastic Activities of Curcumin and Curcumin-Derived Nanoparticles
by Lilian Makgoo and Zukile Mbita

Section 2
Production Technology **103**

Chapter 8 **105**
Sustainable Ginger Production through Integrated Nutrient Management
by N. Divyashree, S. Poojashree, S. Venukumar and Y.C. Vishwanath

Chapter 9 **125**
Ginger Based Agro-Forestry Systems for Livelihood to Rainfed Areas
by Rakesh Chandra Nainwal and Shri Krishna Tewari

Chapter 10 **135**
Cumin (*Cuminium cyminium* L.): A Seed Spice Crop with Adopted Production Technology in Cumin Cultivated Regions
by Navjot Singh Brar, Prakash Mahala, Kartik Sharma, Parmdeep Singh Dhanda, Alpa Yadav, Meenakshi Sharma and Prashant Kaushik

Chapter 11 **145**
Essential Oil of Ginger: Effect of Cultivation and Uses
by Temitayo Esther AdeyeOluwa

Preface

Ginger is a member of the family Zingiberaceae. Ginger was probably initially domesticated by peoples from Maritime Southeast Asia, who brought it with them during the Austronesian expansion (about 5,000 years ago), as they traveled across the Indo-Pacific region all the way to Hawaii. Ancient Greeks and Romans made use of ginger, one of the earliest spices to be sold from Asia and brought to Europe as part of the spice trade. Because of their flavour, the dicot plants belonging to the genus *Asarum*, which are very distantly related to one another, are sometimes referred to as wild ginger. The output of ginger throughout the globe reached 4.1 million tonnes in 2019, with India accounting for 44% of the global total.

Ginger is a natural spice that is used in cuisines worldwide to add a flavorful punch to the food that is being cooked. It also has a long history of use as a traditional herbal remedy for a number of common conditions. Although it is used in traditional medicine and as a dietary supplement, there is insufficient evidence to support the claim that ingesting ginger or its extracts has any positive impact on human health or is a cure for various illnesses. However, there is a good deal of interest in using ginger in therapeutic applications where there is an expectation of clinical results. The therapeutic uses of ginger may be divided into six categories: nausea and vomiting, gastrointestinal function, pain, inflammation, metabolic syndromes, and other symptoms. According to the findings of the vast majority of studies, ginger has been shown to be effective in a variety of settings, such as the treatment of nausea and vomiting during pregnancy (NVP), the improvement of digestive function, the reduction of colorectal cancer risk markers, and anti-inflammatory functions.

This book contains 12 chapters, each of which examines a different aspect of the therapeutic uses of ginger, providing an overview of current advances in the field. The book will be a useful reference manual for those engaged in the study of ginger, including students, academics, and industry professionals. In addition, I hope that it will inspire new research ideas for the future.

Dr. Prashant Kaushik
Universitat Politecnica de Valencia,
Valencia, Spain

CULTA Inc.,
Tokyo, Japan

Section 1

Medicinal Properties

Chapter 1

Introductory Chapter: Curcumin and Its Therapeutic Potency

Rabia Shabir Ahmad, Muhammad Imran, Muhammad Kamran Khan, Muhammad Haseeb Ahmad, Ali Imran and Huda Ateeq

1. Introduction

Turmeric (Curcuma longa Linn) a natural herb, is a Zingiberaceae plant that is grown mostly in subtropical and tropical climates all over the world. It's a product of Southeast Asia, Indonesia, and India [1]. Mostly, it is utilized to give colors and flavors to curry, gravy, and mustards. In the sub-continent, it is utilized to keep teeth and gums healthy. It is also used to cure some serious diseases like hepatitis and certain other liver problems. Turmeric is used as a source of medicine in some nations, like India and China [2]. Turmeric has about 200 biologically active compounds, which are recently becoming the subject of investigation. The active bi-phenolic component of turmeric is "Curcumin" (diferuloylmethane), which has been commonly utilized as a natural medicine to cure various diseases hundreds of years ago [3]. Curcuminoids, which include curcumin (Cur) and two similar substances called desmethoxycurcumin (dMC) and bisdemethoxycurcumin (bDC), are primarily responsible for turmeric's pharmacological effect (BdMC). Curcumin has a lot of benefits, such as anti-inflammatory, antioxidant, chemoprotective, metabolic regulating, immune-modulating, antibacterial, antifungal, antiviral, and anti-depressant qualities, as well as antineoplastic capabilities [4].

2. History and discovery of curcumin

Curcumin can be obtained from C. longa rhizomes, a plant of Zingiberaceae, and it is the main functional compound in turmeric. Curcumin was initially found around 200 years ago when Vogel and Peltier extracted a "yellow substance" from C. longa Linn's rhizomes and termed it curcumin. Milobedeska et al. [5] described it, and Lampe [6], made it for the first time.

3. Therapeutic potential of curcumin

3.1 Curcumin and its anti-inflammatory properties

Curcumin's anti-inflammatory properties are well-known, and several therapeutic investigations had been approved to assess its bio-active properties in a variety of inflammatory situations. Curcumin is well-known for its powerful antioxidant activity as well as strong anti-inflammatory activity. It helps and fights with the body's immune cells against foreign invaders. Curcumin helps to suppress the molecules that cause inflammation in the body tissues. It is effective against "Arthritis", a chronic illness defined by acute joint inflammation that leads to joint damage and disability Curcumin's anti-inflammatory efficacy has largely been studied in individuals with osteoarthritis and rheumatoid arthritis and its anti-arthritic activity has been verified [7, 8].

Various doses of oral curcumin administration formulations (200–2000 mg/day) and its combination with other plant extracts have been studied for a period of (2 weeks–6 months) in which different symptoms of inflammatory, stress, and skin diseases were studied. The results suggested by Dcodhar et al., that curcumin and phenylbutazone had similar effects on morning sickness, swelling of the joints, and walking time in patients with rheumatoid arthritis, but none had an impact on erythrocyte sedimentation rate (ESR), grip strength and joint index [9].

Curcumin is an indispensable component in the treatment of many types of skin diseases and allergies as well, including psoriasis, pruritus, and radiation-induced dermatitis. The research was performed for the effects of oral curcumin administration in psoriasis, a severe inflammatory and hyper-proliferative condition of skin [10] in the form of a tonic or an ointment [11]. The ointment of curcumin lowered serine/threonine-specific protein activity of kinase and the expression of keratinocyte transferrin receptor (TRR), as well as parakeratosis severity and epidermal cell density. A poor response rate was noted in the outcome of oral intake of curcumin in psoriasis patients. Only two patients out of a total of 12 showed signs of improvement [12].

Orally given curcumin, on the other hand, lowered Psoriasis Area and Severity Index (PASI) scores when paired with topical treatments [10]. Clinical research on the effect of curcumin on various ophthalmological problems revealed that this chemical has high effectiveness whether administered by oral ingestion. In addition, individuals with central serous chorioretinopathy saw a considerable improvement after taking oral curcumin [13].

3.2 Curcumin in cardiovascular protection

Curcumin's cardioprotective efficiency has been proven in a growing number of clinical investigations, owing to its anti-atherosclerotic and anti-hyperlipidemic properties. The active components of curcumin help in the regulation of epithelium walls of the blood vessels and help in smooth blood flow. An oral administration of curcumin in doses ranging from 20 to 4000 mg has been shown to improve the blood lipid profile of the patient along with the increase in the levels of antioxidant status [14]. The findings of the study by Baum et al., suggested curcumin intake (4000 mg/day) for 180 days altered TG levels and had no effect on other metabolic parameters like TC, HDL-C, or LDL-C. Higher doses, on the other hand, were shown to be ineffective [15].

3.3 Curcumin and gastrointestinal diseases

The bio-active components of turmeric and curcumin help in the proliferation of healthy gut microbiota. Recently, it is also being used for the treatment of gut dysbiosis (unbalance of gut microbiota that is associated with digestive problems and inflammation) that is also linked with a number of metabolic diseases as well.

Many studies were performed against irritable bowel syndrome (IBS), inflammatory bowel disease (IBD), pancreatitis, helicobacter pylori infections, and ulcers.

However, there is limited literature on pancreatitis and a lot more work in this field is still required. The patients having pancreatitis were treated with 500 mg dosage of curcumin along with piperine and which results in the reduction of erythrocyte malondialdehyde (MDA) levels significantly, indicating curcumin's antioxidant effect [16].

Hanai et al. found that giving patients with ulcerative colitis 1–3 g of oral curcumin along with mesalamine/sulfasalazine for almost 27 weeks, which resulted in a significant drop in ulcers rate [17]. A study conducted by Lang et al. [18] found that individuals having ulcers were subjected to 3 g dosage of curcumin for 30 days, which resulted in much better results than the placebo group. Indeed, 53.8% of the curcumin group reached a reduction of the ulcers in 4th week compared to 0 percent in the placebo group, however, 65.3 percent of the curcumin group obtained clinical response compared to 12.5 percent in the placebo group.

Curcumin also proved efficient for the treatment of H. pylori-related problems. A dosage of 30 mg curcumin was orally taken by a group of examined patients in combination with 100 mg lactoferrin with 20 mg pantoprazole and 600 mg N-acetylcysteine, for only 7 days and resulted in limited overall inflammation severity [19]. Curcumin alone, given at 2100 mg/day for 4 weeks, had a very poor eradication rate and had little effect on inflammatory cytokines [20].

3.4 Curcumin and liver diseases

As curcumin has an antioxidant activity along with other biological properties, it helps the liver to detoxify and helps eliminate the toxins. It can be helpful in those patients that took medicines and drugs for the treatment of ailments i.e. diabetes and other chronic health problems, to reduce the levels of oxidative stress and eliminates toxins along with the smooth flow of the bile duct. According to Kim et al., curcumin supplementation in fermented turmeric powder form was given to subjects and it resulted in high ALT level, Gamma-glutamyl Transferase (GGT), total bilirubin level (TB), and wrapped profiles [21].

3.5 Curcumin and diabetes

Diabetes is another chronic disease in which the human body fails to produce insulin or the body does not respond to the prepared insulin. As a result, sugar concentration in the blood becomes high (hyperglycemia) and it can cause damage to body organs. This condition describes the production and oxidative stress of highly inflammatory cytokines. Curcumin can be useful to treat this condition and its antioxidant stress. Curcumin, first used in the treatment of diabetes, which resulted in blood sugar was treated with 5 grams of turmeric powder and it resulted in the reduction of blood sugar in the affected patients [22].

The impact of Curcumin (*C. Longa)* was inspected in a trial study on post-Prandal, plasma glucose, glycemic index (GI), and insulin levels in a group of healthy patients [23]. After giving the turmeric dosage (6 g), oral glucose-tolerinus test (OGTT) was conducted and there no change was detected in both glycemic index and glucose levels, but significant changes in the levels of insulin were recorded which was the indication that turmeric has an effect on the release of insulin.

3.6 Curcumin and cancer treatment

Multiple factors are involved in the occurrence of cancer. It can be characterized as uncontrolled cell growth that in some cases can spread to other body organs as well. Along with the strong antioxidant activity, curcumin has the ability to fight with cancerous cells as well. Many studies were conducted in order to estimate the positive effects of curcumin in cancer patients. Results from studies showed that many symptoms e.g. multiple myeloma, lesions on the skin, cancers of the neck, head, and orbits, cancers of the lungs, brain, colon, breasts, and prostate were eliminated. It has been confirmed that it can be improved [24]. Mostly, the application of the extracts of turmeric oil also improved symptoms of cancer and improved the condition of lesions. When myeloma patients were given curcumin dosage through oral administration (1–12 g/day), it reduces urinary paraprotein and interpeptide levels of type I collagen and downregulates NFҝB, STAT3, and COX2 [25]. In addition, the use of curcumin has a positive effect on patients with orbital pseudotumor, squamous cell carcinoma of the head, neck, lung, breast, and prostate cancer, with a decrease in the size of the tumor and a reduction in secretory mutants [21]. In colorectal cancer, the patients ingested curcumin in the form of the C3 complex and it significantly reduced M1G and serum TNFα levels, polyps and abnormal crypt foci, and increased apoptosis. Tumor cell count and Bls_2 [26].

4. Conclusion

Turmeric is a commercially important spice (C. longa L., Zingiberaceae) that is widely utilized as a nutritional component in Asia and Western countries. It has long been renowned for its therapeutic and medicinal properties against a wide range of ailments, as reported in traditional, herbal, and other alternative forms of mainstream medicine. The bioactive component of turmeric (curcuminoid) is famous for its abundant pharmacological properties. Researches had shown that active components in turmeric other than curcuminoids are equally useful and have several medicinal potentials. Finally, as autoimmune-related diseases are increasingly associated with cumulative exposure, special attention should also be paid to their prominent potential in both toxic and pollutant exposures.

Author details

Rabia Shabir Ahmad*, Muhammad Imran, Muhammad Kamran Khan, Muhammad Haseeb Ahmad, Ali Imran and Huda Ateeq
Department of Food Science, Government College University Faisalabad, Punjab, Pakistan

*Address all correspondence to: rabiaahmad@gcuf.edu.pk

References

[1] Paramasivam M, Poi R, Banerjee H, Bandyopadhyay A. High-performance thin layer chromatographic method for quantitative determination of curcuminoids in Curcuma longa germplasm. Food Chemistry. 2009;**113**:640-644

[2] Mukerjee A, Vishwanatha JK. Formulation, characterization and evaluation of curcumin-loaded PLGA nanospheres for cancer therapy. Anticancer Research. 2009;**29**:3867-3875

[3] Newman DJ, Cragg GM, Snader KM. Natural products as sources of new drugs over the period 1981–2002. Journal of Natural Products. 2003;**66**:1022-1037

[4] Pandit S, Kim HJ, Kim JE, Jeon JG. Separation of an effective fraction from turmeric against Streptococcus mutans biofilms by the comparison of curcuminoid content and anti-acidogenic activity. Food Chemistry. 2011;**126**:1565-1570

[5] Milobedeska J, Kostanecki S, Lampe V. Zur Kenntnis des curcumins. Berichte der Deutschen Chemischen Gesellschaft. 1910;**43**:2163

[6] Lampe V, Milobedzka J. Studien über curcumin. Berichte der Deutschen Chemischen Gesellschaft. 1913;**46**:2235-2240

[7] Chandran B, Goel A. A randomized, pilot study to assess the efficacy and safety of curcumin in patients with active rheumatoid arthritis. Phytotherapy Research. 2012;**26**:1719-1725

[8] Kertia N, Asdie AH, Rochmah W. Ability of curcuminoid compared to diclofenac sodium in reducing the secretion of cycloxygenase-2 enzyme by synovial fluid's monocytes of patients with osteoarthritis. Acta Medica Indonesiana. 2012;**44**:105-113

[9] Dcodhar S, Sethi, R., Srimal RC. Preliminary study on antirheumatic activity of curcumin (diferuloyl methane). Indian Journal of Medical Research. 2013;(**1**):138

[10] Allegri P, Mastromarino A, Neri P. Management of chronic anterior uveitis relapses: Efficacy of oral phospholipidic curcumin treatment. Long-term follow-up. Clinical Ophthalmology. 2010;**4**:1201

[11] Bahraini P, Rajabi M, Mansouri P, Sarafian G, Chalangari R, Azizian Z. Turmeric tonic as a treatment in scalp psoriasis: A randomized placebo-control clinical trial. Journal of Cosmetic Dermatology. 2018;**17**(3):461-466

[12] Kurd SK, Smith N, VanVoorhees A, Troxel AB, Badmaev V, Seykora JT, et al. Oral curcumin in the treatment of moderate to severe psoriasis vulgaris: A prospective clinical trial. Journal of the American Academy of Dermatology. 2008;**58**:625-631

[13] Mazzolani F, Togni S. Oral administrationofacurcumin-phospholipid delivery system for the treatment of central serous chorioretinopathy: A 12-month follow-up study. Clinical Ophthalmology. 2013;**7**:939

[14] Mohajer A, Ghayour-Mobarhan M, Parizadeh SMR, Tavallaie S, Rajabian M, Sahebkar A. Effects of supplementation with curcuminoids on serum copper and zinc concentrations and superoxide dismutase enzyme activity in obese subjects. Trace Elements and Electrolytes. 2014;**32**:16-21

[15] Baum L, Cheung SK, Mok VC, Lam LVP, Hui E, Lam CW. Curcumin effects on blood lipid profile in a 6-month human study. Pharmacological Research. 2007;**56**:509-514

[16] Durgaprasad S, Pai CG, Alvres JF. A pilot study of the antioxidant effect of curcumin in tropical pancreatitis. Indian Journal of Medical Research. 2005;**122**:315

[17] Hanai H, Iida T, Takeuchi K, Watanabe F, Maruyama Y, Andoh A, et al. Curcumin maintenance therapy for ulcerative colitis: Randomized, multicenter, double-blind, placebo-controlled trial. Clinical Gastroenterology and Hepatology. 2006;**4**:1502-1506

[18] Lang A, Salomon N, Wu JC, Kopylov U, Lahat A, Har-Noy O, et al. Curcumin in combination with mesalamine induces remission in patients with mild-to-moderate ulcerative colitis in a randomized controlled trial. Clinical Gastroenterology and Hepatology. 2015;**13**:1444-1449

[19] Di Mario F, Cavallaro LG, Nouvenne A, Stefani N, Cavestro GM, Iori V, et al. A curcumin-based 1-week triple therapy for eradication of Helicobacter pylori infection: Something to learn from failure? Helicobacter. 2007;**12**:238-243

[20] Koosirirat C, Linpisarn S, Changsom D, Chawansuntati K, Kipasa J. Investigation of the anti-inflammatory effect of Curcuma longa in Helicobacter pylori-infected patients. International Immunopharmacology. 2010;**10**:815-818

[21] Kim SG, Veena MS, Basak SK, Han E, Tajima T, Gjertson DW, et al. Curcumin treatment suppresses IKKβ kinase activity of salivary cells of patients with head and neck cancer: A pilot study. Clinical Cancer Research. 2011;**17**:5953-5961

[22] Srinivasan M. Effect of curcumin on blood sugar as seen in a diabetic subject. Indian Journal of Medical Sciences. 1972;**26**:269-270

[23] Wickenberg J, Ingemansson SL, Hlebowicz J. Effects of Curcuma longa (turmeric) on postprandial plasma glucose and insulin in healthy subjects. Nutrition Journal. 2010;**9**:1-5

[24] Panahi Y, Khalili N, Hosseini MS, Abbasinazari M, Sahebkar A. Lipid-modifying effects of adjunctive therapy with curcuminoids–piperine combination in patients with metabolic syndrome: Results of a randomized controlled trial. Complementary Therapies in Medicine. 2014;**22**:851-857

[25] Rai B, Kaur J, Jacobs R, Singh J. Possible action mechanism for curcumin in pre-cancerous lesions based on serum and salivary markers of oxidative stress. Journal of Oral Science. 2010;**52**:251-256

[26] He ZY, Shi CB, Wen H, Li FL, Wang B, Wang J. Upregulation of p53 expression in patients with colorectal cancer by administration of curcumin. Cancer Investigation. 2011;**29**:208-213

Chapter 2

Anti-Inflammatory and Antioxidant Activities of Ginger

Saja Shareef

Abstract

Ginger (*Zingiber officinale* Rosc.), a member of the Zingiberaceae family, is a medicinal herb utilized for its anti-inflammatory and antioxidant qualities. Ginger's influence on health was discovered due to its high phytochemical content, which includes compounds that eliminate free radicals created by biological systems. Gingerol, shogaol, and other related ginger chemicals limit the body's production of prostaglandins and leukotrienes. They can also suppress the production of pro-inflammatory cytokines, such as IL-1, TNF-, NF-B, and IL-8. According to our knowledge, NF-B activation is associated with a number of inflammatory disorders, including cancer, kidney injury, and Alzheimer's disease.

Keywords: ginger, anti-inflammatory, antioxidant, shogaol, anticancer

1. Introduction

Ginger (*Zingiber officinale* Rosc.) is a member of the Zingiberaceae family. Ginger is native to Southeast Asia and is used in many nations as a spice and condiment to give flavor to cuisine. In addition, ginger rhizome has been utilized in traditional herbal medicine [1]. The pharmacological potential of ginger is ascribed to its rich phytochemistry [2]. Jolad et al. classify fresh ginger into two broad categories: volatiles and non-volatiles. Sesquiterpene and monoterpenoid hydrocarbons are volatiles that give ginger its distinctive scent and flavor. In contrast, non-volatile pungent chemicals include gingerols, shogaols, paradols, and zingerone [3]. The health advantages of ginger are attributed primarily to its phenolic components, such as gingerols and shogaols. Ginger offers many biological actions, including antioxidant, anti-inflammatory, anticancer, renoprotective impact, antinausea, and antiemetic properties, according to accumulated research.

2. Pharmacological activities of ginger (*Zingiber officinale* Rosc.)

2.1 Ginger's antioxidant activity

Overproduction of free radicals, such as reactive oxygen species (ROS), has been shown to play a significant role in the development of numerous chronic diseases [4].

IntechOpen

Ginger's antioxidant activity was assessed *in vitro* using the ferric-reducing antioxidant power (FRAP), 2,2-diphenyl-1-picrylhydrazyl (DPPH), and 2,2′-azinobis-(3-ethylbenzothiazole-6-sulfonic acid) (ABTS) techniques. The number of phenolic compounds in dried ginger was 5,2 times, 1,1 times, and 2,4 times greater than in fresh, stir-fried, and carbonized ginger, respectively. The antioxidant activity of various gingers tended to exhibit the following characteristics: dried ginger > stir-fried ginger > carbonized ginger > and fresh ginger [5].

Moreover, data from FRAP, oxygen radical absorbance capacity, and cellular antioxidant activity experiments revealed that a polyphenol-rich fraction of the dried ginger powder exhibited strong antioxidant activity [6].

Additionally, the kind of extraction solvent may influence the antioxidant activity of ginger. An ethanolic extract of ginger revealed a significant antioxidant capacity and ferric-reducing ability, whereas an aqueous extract exhibited potent free radical scavenging activity and chelating capacity [6]. In human chondrocyte cells, ginger extract exhibited antioxidant properties, with oxidative stress mediated by interleukin-1 [7]. It increased the production of antioxidant enzymes and decreased the production of reactive oxygen species and lipid peroxidation. Ginger extract lowered the level of malondialdehyde (MDA), which is connected to lipid peroxidation, in stressed rat heart homogenates [8].

Via the nuclear factor erythroid 2-related factor 2 (Nrf2) signaling pathway, ginger and its bioactive components (such as 6-Shogaol) displayed antioxidant action [9]. Nrf2 is a redox-sensitive transcription factor that is mostly expressed in metabolic and detoxifying renal organs and protects against oxidative stress in cells. In addition, ginger phenylpropanoids improved Nrf2 activity and increased the levels of glutathione S-transferase P1 (GSTP1) and the Nrf2 antioxidant response element's downstream effector in foreskin fibroblast cells [10]. In a human mesenchymal stem cell model, the effects of ginger oleoresin on ionizing radiation-induced damage were examined. By translocating Nrf2 to the cell nucleus and stimulating the gene expression of HO-1 and NQO1 (nicotinamide adenine dinucleotide phosphate (NADPH) quinone dehydrogenase 1), oleoresin could reduce the level of reactive oxygen species (ROS). In rats with chlorpyrifos-induced oxidative damage, the 6-gingerol-rich fraction from ginger reduced H2O2 and MDA levels, enhanced antioxidant enzyme activity, and increased glutathione [11].

In addition, ginger extract treatment increased serum levels of antioxidants and testosterone and protected rat testes from cyclophosphamide-induced damage.

2.2 Ginger's anti-inflammatory activity

Several studies have demonstrated that ginger and its active ingredients exhibit anti-inflammatory properties that may protect against inflammation-related disorders [12]. Phoshatidylinositol-3-kinase (PI3K), protein kinase B (Akt), and nuclear factor kappa light chain-enhancer of activated B cells (NF-B) were primarily responsible for the anti-inflammatory effects. In human intestinal cell models, 6-Shogaol also showed protective properties against tumor necrosis factor (TNF-)-induced intestinal barrier disruption. It also blocked the overexpression of Claudin-2 and the disassembly of Claudin-1 by inhibiting the PI3K/Akt and NF-B signaling pathways [13]. Moreover, a 6-gingerol-rich fraction reduced an increase in inflammatory markers such as myeloperoxidase, NO, and TNF- in the brain, ovaries, and uterus of chlorpyrifos-treated rats [11].

2.3 Renoprotective effect of ginger

Many studies revealed the effect of ginger on kidney dysfunction. From these studies, it was found ginger significantly protects the renal cells and reduces the severity of tubular damage caused by gentamicin. Ginger was effective as a prophylaxis agent, but it has no curative activity [14]. Interestingly the presence of polyphenols and flavonoids in the *Z. officinale* extract might be responsible for the antioxidant and nephroprotective activities [15]. Furthermore, it was found that injecting ginger (200 mg/kg) in rats induces a considerable decrease in the concentration of urea and creatinine in rat model, improved, restored, and recovered the affected kidney tissue after inducing kidney injury by azathioprine [16].

2.4 Ginger's kidney protective effect

Numerous research has demonstrated ginger's efficacy on renal impairment. Ginger greatly protects renal cells and lowers the severity of tubular damage produced by gentamicin, according to these investigations. Ginger was useful as a preventative, but it lacks curative properties [14]. Intriguingly, the antioxidant and nephroprotective properties of the *Z. officinale* extract may be due to the presence of polyphenols and flavonoids [15]. Moreover, it was discovered that 200 mg/kg of ginger generates a significant drop in the levels of urea and creatinine in rat models, and improves, restores, and recovers damaged kidney tissue induced by azathioprine [16].

2.5 Ginger's anticancer effect

The method by which ginger acts as a cytotoxin against cancer cells is still a matter of debate among scientists. Ginger contains anti-inflammatory and anti-tumorigenic substances, including [6]-gingerol, [6]-shogaol, [6]-paradol, and zerumbone. Ginger and its bioactive compounds inhibit the progression of colorectal, gastric, ovarian, liver, skin, breast, and prostate cancers [17].

2.6 Ginger's antiemetic effect

Ginger (and its compounds) exert its antiemetic effects peripherally, within the gastrointestinal system, by boosting gastric tone and motility via anticholinergic and antiserotonergic activities [18]. Additionally, it is said to enhance gastric emptying. This combination of roles explains ginger's well-known capacity to alleviate symptoms of functional gastrointestinal diseases, such as dyspepsia, abdominal pain, and nausea, which are frequently associated with impaired gastric motility. Three recent studies have explored the effect of ginger on serotonin (5-hydroxytryptamine, 5-HT3, and 5-HT4) and cholinergic (M3) receptor activity; however, the precise mechanism of action of ginger concerning its antiemetic qualities is yet unknown [19].

2.7 Ginger's effect on liver

In a mouse model, dried ginger (*Zingiber officinale*) suppresses inflammation, improves pathological alterations, and lowers INF and IL6 levels. It may potentially cause liver damage. NF-B activation is inhibited to reduce proinflammatory responses, TNF, IL-6, and other inflammatory cytokines levels [20].

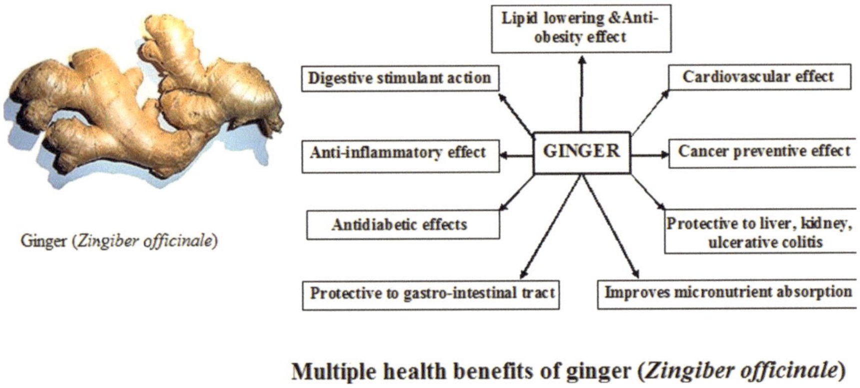

Figure 1.
Multiple health benefits of ginger [28].

2.8 Ginger's effect on diabetes

The consumption of ginger has an impact on lipid profiles, insulin sensitivity, glycemic status, and other metabolic abnormalities in people with type 2 diabetes mellitus (T2DM). Reducing inflammatory substances, such as CRP, IL6, TNF, and others, improves them [21].

It reveals an antagonistic effect against serotonin receptors. Notably, it decreases intestine glucosidase and amylase activity, which reduces glucose absorption [22]. Ginger's neuroprotective effects on the brains of streptozotocin-induced diabetic rats may also result from changes in astrocyte function. Damage response, decreasing the expression of acetylcholinesterase (AchE), and improving the construction of neurons [23].

2.9 Ginger effect on neurological degenerative diseases

Ginger's active component, 6-Shogaol, reduces neuroinflammation and cognitive impairments in animal dementia models. Therefore, it plays a significant role in the amelioration of symptoms in Alzheimer's and other neurological illness patients. It enhances memory in animal models of dementia by suppressing the activity of glial cells and by minimizing memory corruption [24].

In addition, ginger inhibits NF-B,16,17 iNOS, and cyclooxygenase2 activity (COX2) [25]. It protects C57BL/6 mice from inflammation caused by ultraviolet B [26]. Ginger has an inhibitory effect on melanogenesis in B16F10 melanoma cells and can therefore prevent the darkening of the skin (**Figure 1**) [27].

3. Conclusion

Ginger's pharmacological effects are generally known. Anti-oxidant and anti-inflammatory responses allow it to cure a wide variety of disorders. Ginger's anticancer effect is well-documented, and its functional components, gingerols, shogaol, and paradols, are useful chemicals that can prevent various malignancies, metastasis, and prevention of cell-cycle development. Additionally, it enhances the condition of the kidneys and digestive tract.

Author details

Saja Shareef
Al-Esraa University College, Baghdad, Iraq

*Address all correspondence to: saja.majeed@esraa.edu.iq

References

[1] Mashadi NS, Ghiasvand R, Mofid MR. Anti-oxidative and anti-inflammatory effects of ginger in health and physical activity: Review of current evidence. International Journal of Preventive Medicine. 2013;**4**(1):36-42

[2] Stoner GD. Ginger: Is it ready for prime time? Cancer Prevention Research. 2013;**6**(4):257-262

[3] Jolad SD, Lantz RC, Solyom AM, Chen GJ, Bates RB, Timmermann BN. Fresh organically grown ginger (*Zingiber officinale*): Composition and effects on LPS-induced PGE2 production. Phytochemistry. 2004;**65**(13):1937-1954

[4] Poprac P, Jomova K, Simunkova M, Kollar V, Rhodes CJ, Valko M. Targeting free radicals in oxidative stress-related human diseases. Trends in Pharmacological Sciences. 2017;**38**(7):592-607

[5] Li Y, Hong Y, Han Y, Wang Y, Xia L. Chemical characterization and antioxidant activities comparison in fresh, dried, stir-frying and carbonized ginger. Journal of Chromatography, B: Analytical Technologies in the Biomedical and Life Sciences. 2016;**1011**:223-232

[6] Sakulnarmrat K, Srzednicki G, Konczak I. Antioxidant, enzyme inhibitory and antiproliferative activity of polyphenolic-rich fraction of commercial dry ginger powder. International Journal of Food Science and Technology. 2015;**50**(10):2229-2235

[7] Hosseinzadeh A, Bahrampour Juybari K, Fatemi MJ, Kamarul T, Bagheri A, Tekiyehmaroof N, et al. Protective effect of ginger (*Zingiber officinale* Roscoe) extract against oxidative stress and mitochondrial apoptosis induced by interleukin-1β in cultured chondrocytes. Cells, Tissues, Organs. 2017;**204**(5-6):241-250

[8] Ludis M, Carolina RA, Milena FA, Adriana G. Effect of ginger extract on membrane potential changes and AKT activation on a peroxide-induced oxidative stress cell model. Free Radical Biology & Medicine. 2018;**120**:S49

[9] Peng S, Yao J, Liu Y, Duan D, Zhang X, Fang J. Activation of Nrf2 target enzymes conferring protection against oxidative stress in PC12 cells by ginger principal constituent 6-shogaol. Food & Function. 2015;**6**(8):2813-2823

[10] Ji K, Fang L, Zhao H, Li Q, Shi Y, Xu C, et al. Ginger oleoresin alleviated γ-ray irradiation-induced reactive oxygen species via the Nrf2 protective response in human mesenchymal stem cells. Oxidative Medicine and Cellular Longevity. 2017;**2017**

[11] Abolaji AO, Ojo M, Afolabi TT, Arowoogun MD, Nwawolor D, Farombi EO. Protective properties of 6-gingerol-rich fraction from *Zingiber officinale* (Ginger) on chlorpyrifos-induced oxidative damage and inflammation in the brain, ovary and uterus of rats. Chemico-Biological Interactions. 2017;**270**:15-23

[12] Ezzat SM, Ezzat MI, Okba MM, Menze ET, Abdel-Naim AB. The hidden mechanism beyond ginger (*Zingiber officinale* Rosc.) potent in vivo and in vitro anti-inflammatory activity. Journal of Ethnopharmacology. 2018;**214**:113-123

[13] Luettig J, Rosenthal R, Lee IFM, Krug SM, Schulzke JD. The ginger component 6-shogaol prevents

TNF-α-induced barrier loss via inhibition of PI3K/Akt and NF-κB signaling. Molecular Nutrition & Food Research. 2016;**60**(12):2576-2586

[14] Nasri H, Nematbakhsh M, Ghobadi S, Ansari R, Shahinfard N, Rafieian-Kopaei M. Preventive and curative effects of ginger extract against histopathologic changes of gentamicin-induced tubular toxicity in rats. International Journal of Preventive Medicine. 2013;**4**(3):316-321

[15] Shan B, Cai YZ, Sun M, Corke H. Antioxidant capacity of 26 spice extracts and characterization of their phenolic constituents. Journal of Agricultural and Food Chemistry. 2005;**53**(20):7749-7759

[16] Shareef SM, Khaleel RA, Hameed ZE, Alsaraf KM. The protective effect of zingiber officinale l. Extract on kidney tissues and blood factors of kidney functions after the damage caused by azathioprine. Science Pharmaceutical. 2021;**32**(4):78-86

[17] Hung JYU, Hsu YAL, Li C Te, Ko YC, Ni WC, Huang MS, et al. 6-shogaol, an active constituent of dietary ginger, induces autophagy by inhibiting the AKT/mTOR pathway in human non-small cell lung cancer A549 cells. Journal of Agricultural and Food Chemistry 2009;57(20):9809-9816.

[18] Abdel-Aziz H, Windeck T, Ploch M, Verspohl EJ. Mode of action of gingerols and shogaols on 5-HT3 receptors: Binding studies, cation uptake by the receptor channel and contraction of isolated guinea-pig ileum. European Journal of Pharmacology. 2006;**530**(1-2):136-143

[19] Walstab J, Krüger D, Stark T, Hofmann T, Demir IE, Ceyhan GO, et al. Ginger and its pungent constituents non-competitively inhibit activation of human recombinant and native 5-HT3 receptors of enteric neurons. Neurogastroenterology and Motility. 2013;**25**(5):439-448

[20] Li XH, McGrath KC, Nammi S, Heather AK, Roufogalis BD. Attenuation of liver pro-inflammatory responses by Zingiber officinale via inhibition of NF-kappa B activation in high-fat diet-fed rats. Basic & Clinical Pharmacology & Toxicology. 2012;**110**(3):238-244

[21] Mahluji S, Ostadrahimi A, Mobasseri M, Attari VE, Payahoo L. Anti-inflammatory effects of zingiber officinale in type 2 diabetic patients. Advanced Pharmaceutical Bulletin. 2013;**3**(2):273

[22] Goyal RK, Kadnur SV. Beneficial effects of Zingiber officinale on goldthioglucose induced obesity. Fitoterapia. 2006;**77**(3):160-163

[23] El-Akabawy G, El-Kholy W. Neuroprotective effect of Ginger in the brain of streptozotocin-induced diabetic rats. Annals of Anatomy-Anatomischer Anzeiger. 2014;**196**(2):119-128

[24] Moon M, Kim HG, Choi JG, et al. 6-Shogaol, an active constituent of Ginger, attenuates neuroinflammation and cognitive deficits in animal models of dementia. Biochemical and Biophysical Research Communications. 2014;**449**(1):8-13

[25] Shim S, Kim S, Choi D-S, Kwon Y-B, Kwon J. Anti-inflammatory effects of [6]-shogaol: Potential roles of HDAC inhibition and HSP70 induction. Food and Chemical Toxicology. 2011;**49**(11):2734-2740

[26] Guahk G-H, Ha SK, Jung H-S, et al. Zingiber officinale protects HaCaT cells and C57BL/6 mice from ultraviolet B-induced inflammation. Journal of Medicinal Food. 2010;**13**(3):673-680

[27] Huang H-C, Chiu S-H, Chang T-M. Inhibitory effect of [6]-Gingerol on melanogenesis in B16F10 melanoma cells and a possible mechanism of action. Bioscience, Biotechnology, and Biochemistry. 2011;75(6):1067-1072

[28] Srinivasan K. Ginger rhizomes (*Zingiber officinale*): A spice with multiple health beneficial potentials. PharmaNutrition. 2017;5(1):18-28

Chapter 3

Antioxidant Potential of Phytoconstituents with Special Emphasis on Curcumin

Uday Deokate and Mohini Upadhye

Abstract

Various fruits, vegetables, cereal grains, edible macrofungi, microalgae, and medicinal plants are containing phytoconstituents which are considered to be antioxidants. Polyphenols and carotenoids are the two main kinds of antioxidant phytochemicals and they contribute the most to the antioxidant properties of plant and its derivatives are widely employed as antioxidants. Turmeric is a rhizomatous herbaceous perennial plant (*Curcuma longa*) of the ginger family. The medicinal properties of turmeric, the source of curcumin, have been known for thousands of years; however, the ability to determine the exact mechanism(s) of action and to determine the bioactive components have only recently been investigated. Curcumin (1,7-bis(4-hydroxy-3-methoxyphenyl)-1,6-heptadiene-3,5-dione), also called diferuloylmethane, is the main natural polyphenol found in the rhizome of *Curcuma longa* (turmeric) and in others *Curcuma* spp. Curcumin, a polyphenol, has been shown to target multiple signaling molecules while also demonstrating activity at the cellular level, which has helped to support its multiple health benefits such as antioxidant, anti-inflammatory, antimutagenic, antimicrobial and anticancer properties. Curcumin has received worldwide attention for its multiple health benefits, which appear to act primarily through its anti-oxidant and anti-inflammatory mechanisms.

Keywords: curcumin, phytoconsituents, free radicals, antioxidant

1. Introduction

Free radicals are produced during routine cellular metabolic processes. These free radicals are considered as important part of the pathological complications including diabetes mellitus, cardiovascular disorders, neurodegenerative disorders, cancer, cataracts, asthamatic conditions, rheumatoid arthritis, inflammatory conditions, intestinal complications, ischemic and postischemic conditions.

Antioxidants are those substances which at very low concentrations are capable of significantly reducing or preventing the oxidation of the substrates which can be oxidized are called as antioxidants. There is a highly complex system including enzymatic and non-enzymatic systems which is effective in synergistic way with each other, so as to protect the body cells and different organs from the damage caused b free radicals. There are different types such as endogenous antioxidants and exogenous antioxidants such as the diet or various dietary supplements. There are different examples of dietary substances which actually do not scavenge the free radicals but they stimulate the endogenous activity ultimately categorized as antioxidants. An antioxidant which are considered as ideal, should easily absorbed and eliminate the free radicals and can able to cause chelation of redox metal ions at the levels which are considered as physiologically suitable. This should be active at the aqueous and/or in the membrane domains and can result in effective expression of gene at a positive direction. These endogenous antioxidants suppose to be highly potential in maintain the optimum cellular functioning and ultimately systemic healthy conditions and the well being also. But many a times, these endogenous antioxidants are considered as insufficient to support the oxidative or nitrosative stress, so along with it dietary antioxidants are important for maintain the highest cellular functioning. The highly potential enzymatic antioxidants consist of glutathione peroxidase, catalase, superoxide dismutase. Various examples in nonenzymatic antioxidants are vitamin C, vitamin E, thiol antioxidants including lipoic acid, glutathione etc., carotenoids, melatonins, flavanoids and other resembling compounds. One of the mechanisms called as antioxidant network, which involves the regeneration of the original properties of one antioxidant after interaction with other antioxidants. In various diseased conditions, it is reported that there is established link between the raised levels in ROS or RNS and deterioration of actions related to enzymatic or non enzymatic antioxidants [1–3].

2. Enzymatic and non enzymatic antioxidants

2.1 Enzymatic antioxidants

2.1.1 Glutathione peroxidase

The oxidation of the glutathione directed as hydroperoxide which can be considered as hydrogen peroxide or others for example lipid hydroperoxide.

$$\mathrm{ROOH} + 2\mathrm{GSH} \rightarrow \mathrm{GSSG} + \mathrm{H_2O} + ROH \qquad (1)$$

Selenium dependent that is GPx, EC 1.11.1.19 and independent on selenium for example glutathione S transferase, GST, EC 2.5.1.18 are considered as the two different forms. In the humans, reported are four types of se-dependent glutathione peroxidases which are mentioned as addition of two electrons to carry out reduction of peroxides through formation of selenole (SeOH) and these antioxidant properties of the selenoenzymes carries elimination of peroxides as important substrate for the reaction termed as Fenton reaction. Along with the tripeptide glutathione that is GSH, selenium dependent glutathione peroxides will be acting, and it exists in comparatively higher concentrations present in the cells and also catalyses the formation of hydrogen peroxide or the organic peroxide into water or alcohol, while causing oxidation of GSH simultaneously. It is able to compete with that of catalase and hydrogen peroxide as a substrate so, considered as an important source of giving protection against oxidative or the nitrosative stress which at very low levels [4].

2.1.2 Catalase

The first antioxidant enzyme was considered as the Catalase that is EC 1.11.1.6 and it was reported to carry out the conversion of hydrogen peroxide into the water and oxygen as indicated:

$$H_2O_2 \rightarrow 2H_2O + O_2 \tag{2}$$

Amongst all the enzymes, catalase was reported to have the highest rate of turnover, one molecule of the catalase is having capacity to converting around 6 millions of hydrogen peroxides to water and oxygen, each of minute. Some from this catalase is found in all the tissues, majorly activity of catalase is found in liver and erythrocytes [5].

2.1.3 Superoxide dismutase

One from the enzymatic antioxidants that is superoxide dismutase called as EC 1.15.1.1, can reported to be an intracellular enzymatic antioxidant and is responsible to convert superoxide dismutase anions in dioxygen and ions of hydrogen peroxide as:

$$O_2^- + O_2^- + 2H+ \rightarrow H_2O_2 + O_2 \tag{3}$$

This superoxide dismutase, is existing in the form of various isoforms, which majorly different in the natures of the active metal centre, composition of the amino acids, cofactors and other related features. Three different froms related to SOD are available in the human beings for example cytosolic, Zinc- SOD, mitochondrial Mn-SOD, and extracellular SOD. By undergoing successive oxidative and reductive pathways related to transition metal ions at their sites of activity, neutralization of superoxide ions by superoxide dismutase is carried out [6].

2.2 Nonenzymatic antioxidants

2.2.1 Vitamin E

Vitamin E majorly available as fat soluble and it can be found in eight different varieties [7]. A chromanol ring along with the phytyl tail and difference in numbers and position related to methyl groups in these rings are observed in the tocopherols including α, β, γ, and δ. These substances are reported to be lipid soluble with a prominent antioxidant potential. According to the literature, these are found to be more active than that of polyunsaturated fatty acids which are having peroxyl radicals and thus show their action by breaking the lipid peroxidation [8].

The important role of this vitamin is to give protection from the lipid peroxidation and also literature is available to support this, there is synergestic effect of α tocopherol and ascorbic acid, involved in cyclic type reactions. Due to the donation of the hydrogen from α tocopherol to the lipid radical or lipid peroxide radical and it gets converted to α tocopherol radical. Further, this α tocopherol radical can get reduced in the form of α tocopherol due to the ascorbic acid [9].

2.2.2 Vitamin C-ascorbic acid

This considered as one of the antioxidants reported to work in an aqueous medium of the body. In the presence of metal ions it is oxidized into dehydroascorbic acid in an extracellular environment of the body, which is further carried into the cells with help of glucose transporters. It works alone with the antioxidant enzymes and also Vitamin E and carotenoids are termed as its primary partners. From the α tocopherol radicals it forms α tocopherol in the membranes and lipoproteins, vitamin C plays an important role along with vitamin E, ultimately it increases the levels in glutathione in cells, thus protecting the protein thiol groups against oxidative damage. It is present in the reduced form in the cells, due to reaction with glutathione, thus catalyzing protein disulfide isomerase and glutaredoxins. Ascorbic acid is reported as the best potential reducing agent that effectively causes neutralization of ROS for example hydrogen peroxide [10].

2.2.3 Thiol antioxidants

This is a very important intracellular thiol antioxidant which is present in intracellular region is tripeptideglutathione which is also known as GSH. Acoording to the literature, it is considered as a redox buffer of the cell. It is found to be present in the cytosol, mitochondria and nuclei and present in the soluble form of antioxidants in these compartments. The levels of glutathione are important for inducing the postmititic phenotype, which are confirmed based upon the studies on the human fibroblasts, and hence it is reported that the implementation of the depletion of glutathione has very important role during the control in aging at cellular level from human skin.

As this glutathione acts as cofactor for various enzymes which cause detoxifying actions, its participation in the amino acid transport through the plasma membrane, scavenging of hyrdroxyl radical and direct scavenging of the singlet oxygen, regeneration of the active forms of vitamin C and E, has a very potential role for protective action against oxidative or nitrosative stress [11].

Decrease in the glutathione levels are the signals of the oxidative stress which is increased in the ischemic brain disorders, cancer, cardiovascular disorders and decreased concentrations of glutathione are also indicative of both that is type 1 and type 2 diabetes mellitus [12–16].

2.2.4 Thioredoxin

Another potential example of the thiol antioxidants is thioredoxin which is also known as TRX system, which are the types of proteins available both in the mammalian and prokaryotic cells and capable of oxidoreductase activity. It is consisting of disulphide and two cysteins which are redox active with conserved active sites as Cys-Gly-Pro-Cys. This antioxidants consists of two –SH groups which are adjacent forms which are present in the reduction form, then are converted into disulphide units in the oxidized form as TRX after it undergoes redox reactions including multiple proteins in it. The levels of thioredoxin is comparatively lesser than that of GSH, but these may be having functions which are overlapping and in similar compartments related to the stimulation and regulation of various transcription factors. Many of the normal and neoplastic cells, secrets this thioredoxin in the oxidative or nitrosative stress and also in the inflammatory conditions [17, 18].

2.2.5 α-Lipoic acid—1, 2-dithione-3-pentanoic acid

Metal chelating and antiglycation potentials are associated with this third thiol antioxidant which is present as natural substance called as α-Lipoic acid called as ALA. Different than other antioxidants which are either lipid soluble or soluble in aqueous medium, lipoic acid can be considered as active in the aqueous as well as lipid phases. In most of the tissues, lipoic acid is transferred to dihydrolipoic acid that is DHLA through the action of NADH or NADPH, and then it can be readily digested and absorbed too. There are various actions which are related to lipoic acid such as direct termination of the free radicals, chelation of transition metal ions for example copper and iron, glutathione levels observe to be increased, vitamin C levels and prevention of the toxicities which are associated with their loss [19].

Lipoic acid is also capable of crossing the blood brain barrier and thus it can be covered by all central and peripheral regions of nervous system. For the free radicals associated with the oxidative stress, lipid peroxides that is LPO is considered as the biomarker. A sequence of the reactions is initiated after the action of free radicals on the polyunsaturated fatty acids called as PUFA in the biological systems, which ultimately results in the production of the conjugated dienes and lipid peroxidases [20].

2.2.6 N-acetylcysteine

To decrease the conditions related oxidative or nitrosative stress, one of the thiol antioxidant is N-acetylcysteine. It prevents from liver damage related to paracetamol caused in the human beings, causes attenuation in liver damage and also reported to prevent the GSH depletion in the mice [21].

N-acetylcysteine also possesses properties such as metachelation, and is implemented in several clinical conditions. Due to the thio groups present in this N-acetylcysteine, it decreases the free radicals and supplies chelation sites to that for metals. Hence, in the metal poisoning conditions and various diseased conditions, N-acetylcysteine is found to have a potential role as is effective in renovating inbalanced prooxidant and antioxidant conditions.

2.2.7 Melatonin

The melatonin shows its free radical scavenging potential due to donation of electron so as it can release variety of ROS or RNS, containing majorly toxic hydroxyl radicals. Along with that melatonin also increases an enzymes which are considered as antioxidative such as superoxide dismutase, glutathione peroxidase, glutathione reductase, catalase etc [22].

This efficiency related to electron transport chain is reported to be increased and as a result decreased electron leakage and production of the free radicals. Due these highly potential actions of melatonin, it is considered as a very important entity used in treating variety of neurological diseases which are having oxidative damage as a part of the etiology concerned [23].

2.2.8 Carotenoids

One of the potential antioxidants, which are reported to be lipid soluble and containing isoprenoid carbok skeleton is the carotenoids. These are reported to be present in the membranes along with the lipoproteins and a very useful example is

ROS	Neutralizing antioxidants
Hydroxyl radical	Vitamin C, glutathione, Flavonoids, lipoic acid
Superoxide radical	Vitamin C, glutathione, Flavonoids, SOD
Hydrogen peroxide	Vitamin C, glutathione, beta Carotene, vitamin E, CoQ10, Flavonoids, lipoic acid
Lipid peroxides	β-carotene, vitamin E, ubiquinone, flavonoids & Glutathione peroxidase

Table 1.
Various reactive oxygen species and their neutralizing antioxidants.

α- carotene. The efficiency of these antioxidants is considered to be as equivalent as that of α tocopherol as they effectively scavenge singlet oxygen and also result in trapping of the peroxyl radicals at low levels of oxygen pressures (**Table 1**). Those caratenoids show brings are distinguished with the pro-vitamin A activity. As in β- carotene, it indicates presence of two brings on the both the ends related to carbon chain, it possesses highest activity. Vitamin A is also reported to show highly potential antioxidant activity which is independent on that of the oxygen concentration [24].

2.2.9 Flavonoids

One of the highly potential antioxidants which are considered as a part of regular diet and considered as a wide class of plant metabolites having a low molecular weight are Flavanoids. These are considered as the groups of derivatives of benzo-γ- pyrone which are made up of phenolic and the pyrane ring with attached hydroxyl groups are added during normal metabolism. The most important activity of flavanoids is considered as their protective action against that of oxidative or nitrosative stress. The catalytic breakdown of the hydrogen peroxide radicals that is Fentonn chemistry, is resulted due to the scavenging of peroxyl radicals, minimizing lipid peroxidation, chelation of redox active metals. In other certain conditions, flavanoids are reported to indicate the poroxidant activity which is considered to be in direct proportion to the presence of total hydroxyl groups and as per literature also indicated cell signaling modulation [25].

The process called as Oxidation, involves the formation of various intermediates which are reactive and can cause aerobic cell metabolism. Thus it is reported that the cells are having a chance of damage which is protected by this antioxidant systems in the cells. Different pathological conditions such as cancerous conditions, cardiovascular disorders, cataract, diabetes mellitus, gastrointestinal disorders, liver disorders, macular degenerations periodontal disorders, other inflammatory conditions are observed due to loss of balance in the ROS production and defence to antioxidants thus causing oxidative stress as a result of deregulation of the cellular functioning [26].

3. Turmeric containing curcumin as potential phytoconstituent

Curcuma longa, as turmeric is one of the perennial herbs and classified under the family Zingiberaceae. This is cultivated majorly in India and China. Yellow powder of the rhizome from the plant is considered for many of the medicinal purpose. Dried form of *Curcuma longa* that is turmeric used as an ingredient in many of the food preparations. It is known by many other names such as Indian saffron Curcum in the, Arab region, Jianghuang (yellow ginger in Chinese), Haridra (Sanskrit, Ayurvedic), Kyoo or Ukon (Japanese) [27].

Powder of Turmeric is applicable as flavoring and coloring agent in various food preparations. For maintaining oral hygiene, it has been in use from many years [28]. In India and China, turmeric is considered as the choice of treatment for jaundice and other liver problems. This is one of the potential herbs having different pharmacological potentials including anti-protozoal, anti-venom activities, antioxidant, anti-microbial, anti-angiogenic, anti-malarial, anti-inflammatory, anti-proliferative, anti-tumor and anti-aging properties [29–40]. It has also been used to treat parasitic infections, ulcers, skin disorders, immunity related disorders and curing the symptoms of colds and flus [41]. Curcumin (CUR) and two related compounds demethoxy curcumin (DMC) and bisdemethoxycurcumin (BDMC) are considered as important curcuminoids responsible for the pharmacological action of turmeric. The active constituents of present are considered as a mixture of bisdesmethoxycurcumin, curcumin (diferuloylmethane), monodexmethoxycurcumin. Around 90% of curcuminoids are present in turmeric. Apart from this it also consists of proteins, sugars, and resins 0.3–5.4% of curcumin is found to be present in the raw turmeric [42].

Turmeric is containing a mixture of three curcuminoids: curcumin (diferuloylmethane), demethoxycurcumin and bisdemethoxycurcumin, along with that volatile oil containing atlantone, zingiberone and tumerone, proteins resins and sugars. Lipophilic polyphenol that is Curcumin is nearly insoluble in water but is found to be stable in the acidic pH of the stomach [43].

Curcumin is containing phenolic groups due to which it can eliminate free radicals derived from oxygen.

The free radicals such as hydroxyl radical, singlet oxygen, superoxide radical, nitrogen dioxide and NO which can be eliminated by curcumin [44].

Curcumin and its derivatives has the most potential biological effects disease prevention and health promotion including inflammatory, antioxidant and anticancer potential [45].

3.1 Anti-inflammatory properties

Curcumin activity for inflammation after giving oral administration was comparable to that of cortisone or phenylbutazone. *Curcuma longa* after this treatment has potentially reduced inflammatory swelling. This effect can be resulted due to its potential of inhibiting biosynthesis of inflammatory prostaglandins from arachidonic acid and neutrophil function during inflammatory states [46, 47].

3.2 Hepatoprotective activity

Turmeric is having hepatoprotective activity similar to that of silymarin. From studies, it can be concluded that turmeric has hepatoprotective potential in various including carbon tetrachloride (CCl4), acetaminophen (paracetamol) and galactosamine. This hepatoprotective effect is mainly a observed due to the antioxidant activity of turmeric along with its ability to decrease the formation of proinflammatory cytokines. Administration of curcumin is resulted in decrease of liver injury [48–50].

3.3 Anticarcinogenic properties

All three stages of this carcinogenesis-initiation, promotion, and progression are inhibited by curcumin.

During initiation and promotion, curcumin modulates transcription factors controlling phase I and II detoxification of carcinogens; down-regulates proinflammatory cytokines, free radical-activated transcription factors, and arachidonic acid metabolism vicyclooxygenase and lipoxygenase pathways and scavenges free radicals [51, 52].

3.4 Antidiabetic activity

ar-turmerone, ar-turmerone, curcumin, demethoxycurcumin and bisdemethoxycurcumin present in hexane and ehanol extract had reported to stimulate adipocyte differentiation in a dose dependent manner. The results also concluded that turmeric ethanolic extract found to contain curcuminoids and sesquiterpenoids is more strongly hypoglycemic as compared to either sesquiterpenoids or curcuminoids [53].

3.5 Antimicrobial activity

Antibacterial, antifungal, cytotoxic, insecticidal and phytotoxic activity of an ethanolic extract of turmeric was evaluated. The extract showed antifungal activity against Trichophyton longifusus and Microsporum canis and weak antibacterial activity against Staphylococcus aureus. Toxic activity was observed against Lemna minor [54].

3.6 Antidepressant properties

In cases of chronic mild stress that is CMS model, curcumin was studied for antidepressant activity.

Ethanolic extract had indicated increase in sucrose intake as compared to normal control levels, reduction in serum IL-6 and TNF-α and CRF levels in medulla oblongata and serum to limits lower than normal.

Cortisol levels were found to be reduced than the normal levels. Through the inhibition of monoamine oxidize, it had shown its antidepressant activity. It caused reversal of decreased serotonin, dopamine and noradrenalin concentrations and increase in the turnover of serotonin [55, 56].

3.7 Cardiovascular diseases

The protective effect was observed due to lowering of triglyceride and cholesterol levels and decrease in LDL and also due to inhibition of aggregation of platelet [57].

Effect of turmeric extract on cholesterol levels may be due to decreased cholesterol uptake in the intestines and increased conversion of cholesterol to bile acids in the liver. Inhibition of platelet aggregation by C. longa constituents is thought to be via potentiation of prostacyclin synthesis and inhibition of thromboxane synthesis [58].

Oral intake of 500 mg/d of curcumin was given for 7 days and further it resulted in comparable decrease in the serum peroxide levels by 33% and increase in HDL cholesterol by 29% and decrease in level of total serum cholesterol by 12% [59].

3.8 Dyspepsia and gastric ulcer

During a phase II clinical trial for peptic ulcer which was diagnosed with endoscopically in 45 subjects, given with 600 mg curcumin five times daily for 12 weeks, it was observed that ulcers were absent in 48% patients after 8 weeks, and in 76%

patients after 12 weeks. In remaining patients also within 1-2 weeks abdominal pain and other symptoms had decreased significantly [60].

3.9 Irritable bowel syndrome

The most common symptoms of irritable bowel syndrome (IBS) are considered as altered bowel habits, abdominal pain, bloating etc.

After pilot study for eight week of IBS patients, it was found that 53% and 60% reduction in IBS prevalence. In post-study analysis, abdominal pain and discomfort scores were reduced by 22 and 25% [61].

3.10 Inflammatory bowel disease

Ulcerative colitis (UC) and Crohn's disease (CD) are considered as two types of IBD, inflammatory bowel disease. In a pilot study performed in 2005, hematological, erythrocyte sedimentation rate (ESR) and biochemical blood analysis, C-reactive protein (CRP) (the latter two inflammatory indicators), sigmoidoscopy, and biopsy were all performed. The authors from this study concluded that curcumin plus standard therapy was more effective in maintaining remission than placebo plus standard UC treatment [62, 63].

3.11 Neurological disorders

Investigations on animal models for Alzheimer's disease (AD) was indicated a direct effect curcumin in reducing the amyloid pathology of AD.

Results have also shown that curcumin exhibited multiple effects in brain. Curcumin is considered as a future drug of therapy for the treatment of various neurological disorders including tardive dyskinesia, diabetic neuropathy and depression [64].

3.12 Antioxidant potential

The two primary mechanisms as antioxidant and anti-inflammatory which explain the benefits of curcumin have proven for various pharmacological actions. Systemic marker of oxidative stress have been found to be improved due to presence of curcumin. Also, from previous literature, it had proven to increase serum activities of important antioxidants such as superoxide dismutase (SOD).

A recent data and analysis of randomized control studies related to the potential effect of supplementation with purified curcuminoids on various oxidative stress parameters had indicated a significant effect on plasma activities of SOD and catalase, as well as serum concentrations of glutathione peroxidase (GSH) and lipid peroxides. The activity of Curcumin on free radicals has been performed based upon many mechanisms. Many forms of free radicals are scavenged by Curcumin, including reactive oxygen and nitrogen species (ROS and RNS, respectively), and can also it was found to alter the activity of GSH, catalase, and SOD enzymes during the neutralization of free radicals and resulted in inhibition of ROS-generating enzymes such as lipoxygenase/cyclooxygenase and xanthine hydrogenase/oxidase. In addition to this, curcumin is considered as a lipophilic compound that makes it an effective scavenger of peroxyl radicals, hence can be effective as a chain-breaking antioxidant [65–71].

Antioxidant activity of the turmeric is evaluated by performing DPPH radical scavenging activity and FRAP values. Many of the literature suggest that, various extracts of turmeric are having antioxidant activities calculated referring to the DPPH radical-scavenging potential.

In this method, 1 mL of the extract was added to around 1.2 mL of 0.003% DPPH in methanol using at varying concentrations (2.5–80.0 µg/mL). The percentage of DPPH inhibition was then calculated using the equation: % of DPPH inhibition = $[(ADPPH - AS\ ADPPH)] \times 100$, in this ADPPH is the absorbance measured for DPPH in the absence of a sample and AS is the absorbance value of DPPH in the presence of either a sample or the standard. DPPH scavenging activity can be given as concentration of a sample required to reduce absorbance of DPPH by 50% (IC50). This value then graphically determined once after plotting the absorbance that is percentage inhibition of DPPH radicals against the log concentration of DPPH, further determining the slope of the nonlinear regression.

Ferric Reducing Antioxidant Power (FRAP) assay was carried out for confirmation of antioxidant potential of the turmeric. Complex of ferric tripyridyltriazine is reduced into ferrous form which then resulted in an intense blue color observed at low pH range. This colour intensity further confirmed by measuring its absorbance value at 593 nm. 200 μL of the extract solution prepared in varying concentrations from 62.5–1000.0 µg/mL was then added to 1.5 mL of the previously prepared FRAP reagent, then reaction mixture was incubated at 37^0 C for 4 min. FRAP reagent was prepared by adding 10 volumes of 300 mM acetate buffer having pH 3.6 with 1 volume of 10 mM TPTZ solution in 1 volume of 20 mM ferric chloride ($FeCl_3.6H_2O$) and 40 mM hydrochloric acid. The FRAP reagent was then prewarmed to 37°C and was used when freshly prepared. Plotting of the standard curve was then done A using an aqueous solution of ferrous sulfate ($FeSO_4.7H_2O$) (100–1000 µmol), with FRAP values expressed as micromoles of ferrous equivalent (µM Fe [II] per 100 g of sample).

The obtained results from antioxidant studies indicated that the free radical scavenging activity may be due to to the high contents of phenolics and flavonoids having a higher reducing capacity. FRAP assay treats the antioxidants in the sample as reductants in a redox reaction and measures the reducing potential of the test sample. These antioxidants exert their activities by donating electron or hydrogen atoms to the ferric complex which converts to ferrous complex (Fe^{3+} to Fe^{2+} -TPTZ complex), thus breaking the radical chain reaction.

Many major diseases such as liver problem, myocardial infarction, diabetes, cancer are believed to be associated with lipid peroxidation and thus causing major cell damage. Curcuminoids and other polyphenols in turmeric can ameliorate and prevent lipid peroxidation, can stabilize the cell membrane, hence proving its significant role in prevention of atherosclerosis. Inhibitory action of turmeric polyphenols such as curcuminoids on lipid accumulation, oxidation, nitric oxide as well as the formation of inflammatory molecules, nuclear factor-kappa B- (NF-kB-) dependent gene expression, and its activation can thus influence therapeutic potential of turmeric in the treatment of pancreatic, hepatic, cancer and intestinal diseases. The ethanolic extract of turmeric can produce promisable symptomatic relief on external cancerous lesions in human. Along with this, curcumin has resulted to be effective in preventing and treatment of many of the neurodegenerative disorders as a free radical scavenger including Alzheimer's disease. Also after giving short-term supplementation it has proved to reduce hematuria, proteinuria, including systolic blood pressure in patients with relapsed or refractory lupus nephritis. By referring all the literature, Curcumin can be considered as a safe adjuvant therapy. The previous studies had indicated that the high antioxidant properties of turmeric was found to inhibit cellular lipid peroxidation and can also ameliorate other oxidative damage caused by free radicals [72–76].

Thus Turmeric is proven to be an important source of high contents of flavonoids, polyphenols, tannins and ascorbic acid. Curcumin as important phytoconstituent of turmeric varieties is and effective and important antioxidant compound and which can be effective in management of various diseased conditions.

Conflicts of interest

The authors declare that there are no conflicts of interest regarding the publication of this paper.

Author details

Uday Deokate[1] and Mohini Upadhye[2]*

1 Government College of Pharmacy, Aurangabad, Maharashtra, India

2 PES Modern College of Pharmacy (for Ladies), Pune, India

*Address all correspondence to: mohiniketh@rediffmail.com

References

[1] Valko M, Leibfritz D, Moncol J, Cronin MT, Mazur M, Telser J. Free radicals and antioxidants in normal physiological functions and human disease. The International Journal of Biochemistry & Cell Biology. 2007;**39**: 44-84. DOI: 10.1016/j.biocel.2006.07.001

[2] Rahman K. Studies on free radicals, antioxidants, and co-factors. Clinical Interventions in Aging. 2007;**2**:219-236

[3] Kurutas EB. The importance of antioxidants which play the role in cellular response against oxidative/ nitrosative stress: Current state. Nutrition Journal. 2016;**15**(71):1-22

[4] Molavian H, Tonekaboni AM, Kohandel M, Sivaloganathan S. The synergetic coupling among the cellular antioxidants glutathione peroxidase/ peroxiredoxin and other antioxidants and its effect on the concentration of H_2O_2. Scientific Reports. 2015;**5**:1-8

[5] Sies H. Oxidative stress: A concept in redox biology and medicine. Redox Biology. 2015;**4**:180-183

[6] Sheng Y, Abreu IA, Cabelli DE, Maroney MJ, Miller A, Teixeira M, et al. Superoxide dismutases and superoxide reductases. Chemical Reviews. 2014;**114**: 3854-3918

[7] Schmölz L, Birringer M, Lorkowski S, Wallert M. Complexity of vitamin E metabolism. World Journal of Biological Chemistry. 2016;**7**:14-43

[8] Singh U, Devaraj S, Jialal I, Vitamın E. oxidative stress, and inflammation. Annual Review of Nutrition. 2005;**25**:151-174

[9] Wang X, Quinn PJ. The location and function of vitamin E in membranes (review). Molecular Membrane Biology. 2000;**17**:143-156

[10] Kojo S. Vitamin C: Basic metabolism and its function as an index of oxidative stress. Current Medicinal Chemistry. 2004;**11**:1041-1064

[11] Birk J, Meyer M, Aller I, Hansen HG, Odermatt A, Dick TP, et al. Endoplasmic reticulum: Reduced and oxidizedglutathione revisited. Journal of Cell Science. 2013;**126**:1604-1617

[12] Bharath S, Hsu M, Kaur D, Rajagopalan S, Andersen JK. Glutathione, iron and Parkinson's disease. Biochemical Pharmacology. 2002;**64**:1037-1048

[13] Warner DS, Sheng H, Batinic'-Haberle I. Oxidants, antioxidants and theischemic brain. The Journal of Experimental Biology. 2004;**207**:3221-3231

[14] Pastore A, Federici G, Bertini E, Piemonte F. Analysis of glutathione: Implication in redox and detoxification. Clinica Chimica Acta. 2003;**333**:19-39

[15] Mills BJ, Weiss MM, Lang CA, Liu MC, Ziegler C. Blood glutathione and cysteinechanges in cardiovascular disease. The Journal of Laboratory and Clinical Medicine. 2000;**135**:396-402

[16] Navarro J, Obrador E, Carretero J, Petschen I, Aviñó J, Perez P, et al. Changes in glutathione status and the antioxidant system in blood and in cancer cells associate with tumour growth in vivo. Free Radical Biology & Medicine. 1999;**26**:410-418

[17] Rubartelli A, Bonifaci N, Sitia R. High rates of thioredoxin secretion correlate with growth arrest in hepatoma cells. Cancer Research. 1995;**55**:675-680

[18] Tonissen KF, Di Trapani G. Thioredoxin system inhibitors as mediators of apoptosis for cancer therapy. Molecular Nutrition & Food Research. 2009;**53**:87-103

[19] Senoglu M, Nacitarhan V, Kurutas EB, Senoglu N, Altun I, Atli Y, et al. Intraperitoneal alpha-lipoic acid to prevent neural damage after crush injury to the rat sciatic nerve. Journal of Brachial Plexus and Peripheral Nerve Injury. 2009;**4**:22-27

[20] Belge Kurutas E, Inanc Guler F, Kilinc M. Free radicals. Archives of Medical Review Journal. 2004;**13**:120-132

[21] Knight TR, Fariss MW, Farhood A, Jaeschke H. Role of lipid peroxidation as a mechanism of liver injury after acetaminophen overdose in mice. Toxicological Sciences. 2003;**6**:229-236

[22] Cetinkaya A, Bulbuloglu E, Belge Kurutas E, Ciralik H, Kantarceken B, Buyukbese MA. Beneficial effects of n-acetylcysteine on acetic acid-induced colitis in rats. The Tohoku Journal of Experimental Medicine. 2005;**206**:131-139

[23] Reiter RJ, Acuña-Castroviejo D, Tan DX, Burkhardt S. Free radical-mediated molecular damage. Mechanisms for the protective actions of melatonin in the central nervous system. Annals of the New York Academy of Sciences. 2001;**939**:200-215

[24] Fiedor J, Burda K. Potential role of carotenoids as antioxidants in human health and disease. Nutrients. 2014;**6**:466-488

[25] Rice-Evans CA, Miller NJ, Paganga G. Structure–antioxidant activity relationships of flavonoids and phenolic acids. Free Radical Biology & Medicine. 1996;**20**:933-956

[26] Homas CE, Kalyanraman B. Oxygen Radicals and the Disease Process. Netherlands: Harwood Academic Publishers; 1997

[27] Goel A, Kunnumakkara AB, Aggarwal BB. Curcumin as "Curecumin": From kitchen to clinic. Biochemical Pharmacology. 2008;**75**(4):787-809

[28] Chaturvedi TP. Uses of turmeric in dentistry: An update. Indian Journal of Dental Research. 2009;**20**:107e109

[29] Mukerjee A, Vishwanatha JK. Formulation, characterization and evaluation of curcumin-loaded PLGA nanospheres for cancer therapy. Anticancer Research. 2009;**29**:3867e3875

[30] Perko T, Ravber M, Knez Z, Skerget M. Isolation, characterization and formulation of curcuminoids and in vitro release study of the encapsulated particles. Journal of Supercritical Fluids. 2015;**103**:48e54

[31] Kalpravidh RW, Siritanaratkul N, Insain P, et al. Improvement in oxidative stress and antioxidant parameters in b-thalassemia/Hb E patients treated with curcuminoids. Clinical Biochemistry. 2010;**43**:424e429

[32] Changtam C, de Koning HP, Ibrahim H, Sajid MS, Gould MK, Suksamrarn A. Curcuminoid analogs with potent activity against Trypanosoma and Leishmania species. European Journal of Medicinal Chemistry. 2010a;**45**:941e956

[33] Lim HS, Park SH, Ghafoor K, Hwang SY, Park J. Quality and antioxidant propertiesof bread containing turmeric (*Curcuma longa* L.) cultivated in SouthKorea. Food Chemistry. 2011;**124**:1577e1582

[34] Peret-Almeida L, Cherubino APF, Alves RJ, Dufosse L, Gloria MBA.

Separation and determination of the physico-chemical characteristics of curcumin, demethoxy curcumin and bisdemethoxycurcumin. Food Research International. 2005;**38**:1039e1044

[35] Aditya NP, Chimote G, Gunalan K, Banerjee R, Patankar S, Madhusudhan B. Curcuminoids-loaded liposomes in combination with arteether protects against Plasmodium berghei infection in mice. Experimental Parasitology. 2012;**131**:292e299

[36] Khan MA, El-Khatib R, Rainsford KD, Whitehouse MW. Synthesis and antiinflammatory properties of some aromatic and heterocyclic aromatic curcuminoids. Bioorganic Chemistry. 2012;**40**:30e38

[37] Yue GGL, Chan BCL, Hon P, et al. Immunostimulatory activities of polysaccharide extract isolated from Curcuma longa. International Journal of Biological Macromolecules. 2010a;**47**:342e347

[38] Tapal A, Tiku PK. Complexation of curcumin with soy protein isolate and its implications on solubility and stability of curcumin. Food Chemistry. 2012;**130**:960e965

[39] Panahi Y, Saadat A, Beiraghdar F, Nouzari SMH, Jalalian HR, Sahebkar A. Antioxidant effects of bioavailability-enhanced curcuminoids in patients with solid tumors: A randomized double-blind placebo-controlled trial. Journal of Functional Foods. 2014;**6**:615e622

[40] Zhan PY, Zeng XH, Zhang HM, Li HH. High-efficient column chromatographicextraction of curcumin from *Curcuma longa*. Food Chemistry. 2011;**129**:700e703

[41] Siviero A, Gallo E, Maggini V, et al. Curcumin, a golden spice with a low bioavailability. J Herb Med. 2015;**5**:57e70

[42] Heath DD, Khwaja F, Rock CL. Curcumin content of turmeric and curry powders. The FASEB Journal. 2004;**18**:A125

[43] Wang YJ, Pan MH, Cheng AL, Lin LI, Ho YS, Hsieh CY, et al. Stability of curcumin in buffer solutions and characterization of its degradation products. Journal of Pharmaceutical and Biomedical Analysis. 1997;**15**:1867-1876

[44] Sreejayan N, Rao MNA, Priyadarsini KI, Devasagayam TP. Inhibition of radiation induced lipid peroxidation by curcumin. International Journal of Pharmaceutics. 1997;**151**: 127-130

[45] Xu X-Y, Meng X, Li S, Gan R-Y, Li Y, Li H-B. Bioactivity, health benefits, and related molecular mechanisms of curcumin: Current progress, challenges, and perspectives. Nutrients. 2018;**10**:1553. DOI: 10.3390/nu10101553

[46] Cronin JR. Curcumin: Old spice is a new medicine. Journal of Alternative & Complementary Therapies. 2003;**9**(1): 34-38

[47] Bundy R, Walker AF, Middleton RW, Booth J. Turmeric extract may improve irritable bowel syndrome symptomology in otherwise healthy adults: A pilot study. Journal of Alternative and Complementary Medicine. 2004;**10**:1015-1018

[48] Ruby J, Kuttan G, Babu KD, Rajashekharan KN, Kuttan R. Antitumorand oxidant activity of natural curcuminoids. Cancer Letters. 1995;**94**:79-83

[49] Rao CV, Desai D, Rivenson A, Simi B, Amin S, Reddy BS. Chemoprevention of colon carcinogenesis by phenylethyl-3-methylcaffeate. Cancer Research. 1995;**55**(11):2310-2315

[50] Park EJ, Jeon CH, Ko G, Kim J, Sohn DH. Protective effect of curcumin in rat liver injury induced by carbon tetrachloride. Journal of Pharmacy and Pharmacology. 2000;**52**:437-440

[51] Garg R, Gupta S, Maru GB. Dietary curcumin modulates transcription a regulators of phase I and phase II enzymes in benzo[a]pyrene-treatedmice: Mechanism of its anti-initiating action. Carcinogenesis. 2008;**29**:1022-1032

[52] Hong J, Bose M, Ju J, Ryu JH, Chen X, Sang S, et al. Modulation of arachidonic acid metabolism by curcumin and related beta-diketone derivatives: Effects of cytosolic phospholipase A(2), cyclooxygenases and 5-liposygenase. Carcinogenesis. 2004;**25**:1671-1679

[53] Nishiyama T, Mae T, Kishida H, Tsukagawa M, Mimaki Y, Kuroda M, et al. Curcuminoids and sesquiterpenoids in turmeric (*Curcuma longa* L.) suppress an increase in blood glucose level in type 2 diabetic KKAy mice. Journal of Agricultural and Food Chemistry. 2005;**53**(4):959-963

[54] Khattak S, Saeedur R, Ullah Shah H, Ahmad W, Ahmad M. Biological effects of indigenous medicinal plants Curcuma longa and Alpinia galanga. Fitoterapia. 2005;**76**(2):254-257

[55] Yu ZF, Kong LD, Chen Y. Antidepressant activity of aqueous extracts of *Curcuma longa* in mice. Journal of Ethnopharmacology. 2002;**83**(1-2): 161-165

[56] Xia X, Cheng G, Pan Y, Xia ZH, Kong LD. Behavioral, neurochemical and neuroendocrine effects of the ethanolic extract from *Curcuma longa* L. in the mouse forced swimming test. Journal of Ethnopharmacology. 2007;**110**(2):356-363

[57] Ramirez-Tortosa MC, Mesa MD, Aguilera MC, Quiles JL, Baro L, Ramirez-Tortosa CL, et al. Oral administration of turmeric extract inhibits LDL oxidation and has hypocholesterolemic effects in rabbits with experimental atherosclerosis. Atherosclerosis. 1999;**147**:371-378

[58] Srivastava R. Inhibition of neutrophil response by curcumin. Agents and Actions. 1989;**28**:298-303

[59] Soni KB, Rajan A, Kuttan R. Reversal of aflatoxin induced liver damage by turmeric and curcumin. Cancer Letters. 1992;**66**:115-121

[60] Prucksunand C, Indrasukhsri B, Leethochawalit M, Hungspreugs K. Phase II clinical trial on effect of the long turmeric (Curcuma longa Linn) on healing of peptic ulcer. The Southeast Asian Journal of Tropical Medicine and Public Health. 2001;**32**:208-215

[61] Barbara G, De Giorgio R, Stanghellini V, Cremon C, Corinaldesi R. A role for inflammation in irritable bowel syndrome? Gut. 2002;**51**(1):i41-i44

[62] Holt PR, Katz S, Kirshoff R. Curcumin therapy in inflammatory bowel disease: A pilot study. Digestive Diseases and Sciences. 2005;**50**:2191-2193

[63] Hanai H, Iida T, Takeuchi K, et al. Curcumin maintenance therapy for ulcerative colitis: Randomized, multicenter, double-blind, placebo controlled trial. Clinical Gastroenterology and Hepatology. 2006;**4**:1502-1506

[64] Kulkarni SK, Dhir A. An overview of curcumin in neurological disorders. Indian Journal of Pharmaceutical Sciences. 2010;**72**(2):149-154

[65] Do QD, Angkawijaya AE, Tran-Nguyen PL, et al. Effect of

extraction solvent on total phenol content, total flavonoid content, and antioxidant activity of Limnophila aromatica. Journal of Food and Drug Analysis. 2014;**22**(3):296-302

[66] Moure A, Cruz JM, Franco D, et al. Natural antioxidants from residual sources. Food Chemistry. 2001;**72**(2):145-171

[67] Cook NC, Samman S. Flavonoids—chemistry, metabolism, cardioprotective effects, and dietary sources. Journal of Nutritional Biochemistry. 1996;**7**(2):66-76

[68] Ghasemzadeh A, Azarifar M, Soroodi O, Jaafar HZ. Flavonoid compounds and their antioxidant activity in extract of some tropical plants. Journal of Medicinal Plants Research. 2012;**6**(13):2639-2643

[69] Sumazian Y, Syahida A, Hakiman M, Maziah M. Antioxidant activities, flavonoids, ascorbic acid and phenolic contents of Malaysian vegetables. Journal of Medicinal Plants Research. 2010;**4**(10):881-890

[70] Tilak JC, Banerjee M, Mohan H, Devasagayam TPA. Antioxidant availability of turmeric in relation to its medicinal and culinary uses. Phytotherapy Research. 2004;**18**(10):798-804

[71] Krup V, Prakash L, Harini A. Pharmacological activities of turmeric (curcuma longa linn): A review. Journal of Homeopathy Ayurvedic Medicine. 2013;**2**(4):133

[72] Yao LH, Jiang YM, Shi J, et al. Flavonoids in food and their health benefits. Plant Foods for Human Nutrition. 2004;**59**(3):113-122

[73] Borra SK, Gurumurthy P, Mahendra J. Antioxidant and free radical scavenging activity of curcumin determined by using different in vitro and ex vivo models. Journal of Medicinal Plants Research. 2013;**7**(36):2680-2690

[74] Bengmark S, Mesa MD, Gil A. Plant-derived health: The effects of turmeric and curcuminoids. Nutricion Hospitalaria. 2009;**24**(3):273-281

[75] Kuttan R, Sudheeran PC, Josph CD. Turmeric and curcumin as topical agents in cancer therapy. Tumori. 1987;**73**(1):29-31

[76] Mishra S, Palanivelu K. The effect of curcumin (turmeric) on Alzheimer's disease: An overview. Annals of Indian Academy of Neurology. 2008;**11**(1):13-19

Chapter 4

Pharmacological Properties of Ginger Combinations

Douglas Ongeri Ochora

Abstract

Ginger (*Zingiber officinale*) is a widely cultivated plant due to its pharmacological properties and its use as a spice. The plant species enjoys a good reputation in most regions of the world mainly because of its anti-inflammatory, antitumor, and antioxidant activities. To enhance these pharmacological properties, ginger is mostly used in combination. Drug combination therapy is also a worthwhile strategy for the prevention of various diseases. Therefore, the current chapter concerted on pharmacological activities of ginger combinations. The plant species has been combined with other plant extracts, pure compounds, and approved drugs for antimicrobial, antioxidant, anticancer, antidiabetic, and antidepressant activities and also in herbal tea formulations. Most of these activities showed synergism with 50% inhibition concertation (IC_{50}) values of less than 1. The highest activity was observed when ginger phytochemicals, shogaol, and gingerol derivatives were combined against prostate cancer cell lines with an IC_{50} value of 0.03. Interaction of different phytochemicals in ginger with other phytochemicals when used in combination account for the reported synergism. The observed synergism in most combinations depicts a potential use of ginger combinations in treatment and prevention of various diseases and disease conditions.

Keywords: combination, ginger, pharmacological, synergism, *Zingiber officinale*

1. Introduction

Most people are exposed to various diseases with minimal opportunities for conventional means of treatment. The people, therefore, rely on traditional herbal medicine [1]. This has made natural products remain a major source of remedies in traditional medicine [2]. Sick people are, therefore, likely to take herbal medication that is easily available before seeking treatment in the orthodox health sector for the recommended form of treatment. This leads to herbal-drug interactions [3]. This has necessitated studies of various medicinal plants when used in combination, which will act as makers for herbal formulations.

Ginger, *Zingiber officinale* Roscoe, belongs to the genus *Zingiber* Mill. which has 149 plant species under the family Zingiberaceae [3]. The plant species are native to Asia but it is currently cultivated in various parts of the world, especially in the tropics [2]. This is because of its wide dietary use as a spice and has long history of medicinal value [4]. The underground stem (rhizome) is the most widely used

IntechOpen

part of ginger. Preference for the rhizome is attributed to the fact that most phytochemicals occur in the rhizome which give the plant its dietary and medicinal benefits [1]. These benefits are enhanced when the plant species being used in combination.

Ginger rhizome has been prescribed for the treatment of various diseases [5] in most traditional and complementary systems of treatment such as homeopathy, Ayurveda, Chinese, Unani Tibb, and Siddha [6]. The plant has been used as an antioxidant [7], antibacterial [8], anticancer [9], anti-inflammation [4], antidepressant [10], regulation of blood sugar level, mensural irregularity [3], treatment of nausea and vomiting, post-operative nausea and vomiting [11], mitigation of rheumatoid arthritis/osteoarthritis/joint and muscle pain, and many other medicinal uses [3]. Studies have been done to determine the activities of ginger when used singly and in combination.

To enhance the medicinal activity of ginger, the rhizome of the plant has been used in combination with other plant extracts, pure compounds, and approved drug in the treatment of various diseases in traditional medicine and in herbal tea formulations, considering tea is the second most widely consumed after water [12]. Phytochemicals present in ginger, especially polyphenols, provide ginger herbal teas with various medicinal properties [13] when combined with phytochemicals from other plants especially polyphenolic compounds [14–16].

The long period and cost incurred in the isolation of pure compounds from plants in drug discovery for conventional medicine buttresses the importance of drug combination studies in phytomedicine [17]. Unlike the use of pharmaceutical drugs whose activity is based on a single active ingredient, in combinations, numerous phytochemicals act collectively [14, 18]. Most of these combinations have shown synergism in prevention and treatment of various diseases and disease conditions. The current chapter, therefore, explored the pharmacological benefits of *Zingiber officinale* when used in combination.

2. Discussion

2.1 Determination of combination activities

The mode of interaction between plant extracts or pure compounds or standard approved drugs is determined using fractional inhibition or interaction factor. Fractional inhibition is used to determine the pharmacological activities of extract-drug combinations, and drug-drug combinations [19]. Fractional inhibition (FI) or combination index (CI) is the inhibition that is attributable to each of the extracts, compounds, or drugs in the combination. Pharmacological activity for each combination is calculated to give a combination index (CI) or 50% fractional inhibition concentration (FIC_{50}) using 50% inhibition concertation (IC_{50}). These CI values are grouped into synergism ($CI < 1$), additivity ($CI = 1$), and antagonism ($CI > 1$), see **Figure 1**.

The Sum CI values are used to generate scatter plots (isobolographs) that explain the potency of a given combination by plotting at least all the triplicate assays on one plane. These isobole curves are used to express the activities based on the dose-response of single components when used alone and in combination at different concentrations [20]. Sum FIC_{50} for each extract-drug, and drug-drug combination ratios are also determined using the equation below [4]:

$$SUM\ CI = \frac{CI_{50}\ \text{of A in combination}}{CI_{50}\ \text{of A alone}} + \frac{CI_{50}\ \text{of B in combination}}{CI_{50}\ \text{of B alone}} \quad (1)$$

Interaction factor (IF) also explains mode of interactions between components in extract-drug and drug-drug combinations. It is also based on isobole graphs and grouped just like fractional inhibition concentration; synergism (IF < 1), additivity (IF = 1), and antagonism (IF > 1) [14, 20].

The IF is calculated as shown below:

$$\text{IF} = AM / AT \quad (2)$$

AM = am measure of activity of a mixture of samples

AT = theoretically calculated activity of the mixture that is based on the dose-response of single components in the combination at different concentrations.

2.2 Antimicrobial activities of ginger combinations

Ginger has been used widely against microbial activities [21]. When used alone, extracts of ginger showed antimicrobial activity with a mean inhibition zone, a 50% microbial inhibition (MIC_{50}) value of 11.72 ± 0.62 mm and lemon (*Citrus limon* L.) showed MIC_{50} value of 27.89 ± 1.12 mm against *Streptococcus mutans*. The activity of ginger when used in combination with lemon was revealed to have a mean inhibition zone with an MIC_{50} value of 17.94 ± 0.46 mm [22]. This study reveals that the anti-microbial activities of ginger are enhanced when used in combination with ginger *C. limon*. Therefore *C. limon* potentiates ginger against *S. mutans*.

Synergism was observed when ginger was combined with barley (*Hordeum vulgare* L.) against *Staphylococcus aureus* and *Aeromonas hydrophila* [23]. Similar synergism was also

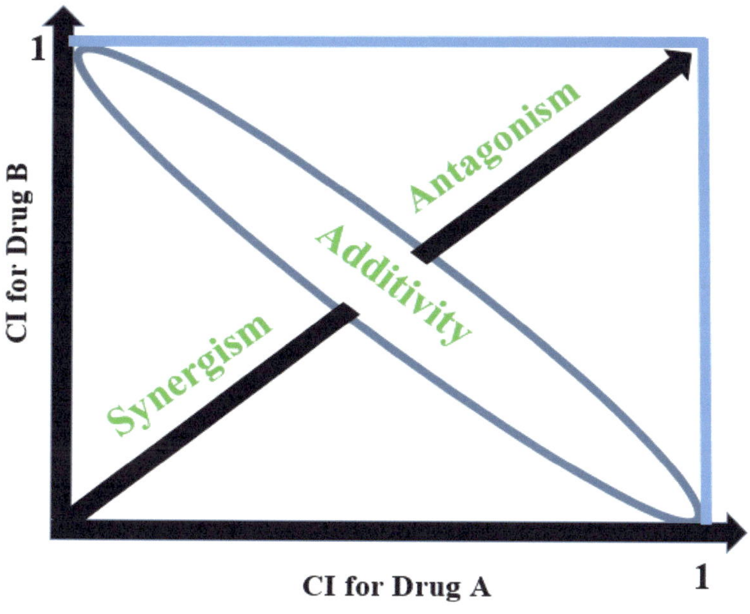

Figure 1.
CI values are grouped into; synergism (CI < 1), additivity (CI = 1), and antagonism (CI > 1).

observed when ginger was combined with pomegranate (*Punica granatum* L.) against *A. hydrophila*. The synergism resulted in complete inhibition of *A. hydrophila.* Combinations of ginger and licorice (*Glycyrrihza glabra* L.), and ginger and sage (*Salvia officinalis* L) tested against *A. hydrophila* and *Escherichia coli* both showed synergism [23]. Combined activity of a plethora of phytochemicals in ginger extracts combined with other plant extracts is responsible for the observed synergism when tested for antimicrobial activities. These studies indicate the potential use of ginger in combination with other plant extracts against microbial diseases.

In another study, methanol extracts from ginger showed antibacterial activities against 32 *S. aureus* strains with an MIC_{90} value of 3.56 mg/mL when used alone. Synergism was observed with CI_{90} values of less than 0.9 when two standard antibacterial drugs, tetracycline and netilmicin were each combined with ginger extracts [24]. This shows the potential use of ginger extracts in combination with tetracycline and netilmicin in controlling antimicrobial diseases and reduction of antimicrobial drug resistance. In the same study, additivity and antagonism were observed when ginger was combined with chloramphenicol, erythromycin, gentamycin, vancomycin, ampicillin, cefoxitin, cotrimoxazole, and ofloxacin, all tested against *S. aureus* [24]. In another comparative study, methanol and water leaves and rhizome ginger extracts were shown to be more effective against *S. aureus* and *S. pyogenes* than three standard antibiotics: chloramphenicol, ampicillin, and tetracycline. This depicts the possible use of ginger alongside standard antibiotics in the treatment of microbial diseases [12].

One of the major burdens of disease control is the emergence of drug-resistant strains. When two drugs that have different mechanisms of action are used in combination the likelihood of the emergence of drug-resistant strains is reduced. The pace of emergence of drug resistance is also reduced since a lower amount of each drug is used in combination treatment [19]. Therefore, these studies underpin the use of ginger combinations against antimicrobial diseases. The potential use of ginger in combination with other plants used against microbial diseases and a combination of ginger with standard antibiotic drugs is also depicted in overcoming antimicrobial drug resistance.

2.3 Antioxidant activities of ginger combinations

Ginger has been reported to have high phenolic content and high antioxidant activity [7]. The combination of ginger and coffee (*Coffea arabica* L.) has shown antioxidant synergistic effects. A combination of ethanol extracts of coffee and ginger showed the highest 2,2-dipheny-1-picryl-hydrazyl-hydrate (DPPH) radical scavenging activity than when used singly. Similarly, combined ginger and tea aqueous extracts showed the highest ABTS radical scavenging activity [13]. This shows that the antioxidant activities of ginger and tea are higher when used in combination than when used singly.

Juice extracts of ginger, kesum (*Polygonum minus* H.), and turmeric (*Curcuma longa* L.) without using any solvent showed were tested for antioxidant activities using DPPH radical scavenging assay and ferric-reducing antioxidant power (FRAP) assay [25]. When tested singly for FRAP ginger showed antioxidant activities of 26.2 μmol/g, kesum 46.3 μmol/g, and turmeric 23.3 μmol/g (μmol of Fe II/g of extracts). When the three plant species were combined at a ratio of 1:1:1, they had an antioxidant activity of 23.1 μmol/g which shows that kesum and turmeric juice extracts potentiate antioxidant activities of ginger. When ginger was combined separately with kesum (1:1) and turmeric (1:1) for FRAP, antagonism was displayed with CI_{50} values of greater than 1. For DPPH radical scavenging activities, juice extracts of ginger, kesum, and turmeric showed percentage antioxidant activities of 79.0%, 82.6%, and 64.6%, respectively, when tested singly. When ginger was combined separately with extracts of both kesum (1:1) and turmeric (1:1) and when they were all combined (1:1:1) for DPPH radical scavenging activities, they showed additivity with CI_{50} values of 1 [25].

In another study, the combination of ginger, garlic (*Allium sativum* L.), and cayenne pepper (*Capsicum fructensces* L.) extract at a ratio of 1:1:1 showed synergism with CI_{50} values of less than 1. All extracts showed higher antioxidant activities than when these three natural spices are used alone [26] indicating that the extracts potentiate each other's antioxidant activities. This is probably because the many phytochemicals in the combination increase the superoxide dismutase (SOD), glutathione peroxidase (GPx), and glutathione reductase (GRed) enzymes. These enzymes catalyze the conversion of reactive oxygen species to harmless species [26].

Ethanol extracts of ginger and secang wood (*Caesalpinia sappan* L.) were tested for antioxidant activities using DPPH method both in combination and singly [27]. The concertation of antioxidant compounds was highest in a combination of scang and ginger (2:1) with an IC_{50} value of 90,14; absorbance was measured using UV. When tested singly, the IC_{50} for scang and ginger was 54,53 and 197,74, respectively [27]. In another similar study, ginger combination with eucalyptus honey (*Eucalyptus globulus* Labill.) portrayed synergistic activities with CI_{50} values of less than 1 in which antioxidant activities were increased where H6G5 displayed 60.0% and 90.0% using DPPH method [28].

2.4 Anticancer activities of ginger combinations

Shogaol and gingerol compounds and their derivatives isolated from ginger have shown activity against different types of cancers; breast cancer, lung cancer, cervical cancer, prostate cancer, liver cancer, blood cancer, and colorectal cancer [29]. Combinations of pure ginger phytochemicals, 6-gingerol (6G), 8-gingerol (8G), 10-gingerol (10G), and 6-shogoal (6S) have shown strong synergistic antiproliferative activities against prostate cancer cell lines with CI_{50} values ranging between 0.03 and 0.88 [15]. The highest synergistic activities were observed in 6G + 8G, 8G + 10G, 10G + 6G, and 6S + 10G combinations with CI_{50} values below 0.4 when tested against human prostate cancer cell lines [15]. In another follow-up study, a combination of 6-gingerol, a pure compound isolated from a Chinese *Tongling* White Ginger, with standard drugs depicted synergistic anticancer activities when tested against human cervical adenocarcinoma cells [30]. The combination of 6-gingerol with 5-FU (inhibition of S-phase of cell cycle and PI3K signaling pathway inhibitor) and Ptx (inhibition of G2/M-phase of cell cycle and mTOR inhibitor) at different concentrations showed synergism after 48 h treatment on Hela cells. A combination of 6-gingerol (45 μM) with Ptx (0.36 μM) inhibited 83.2% growth of the treated cells and a combination of 6-gingerol (45 μM) with 5-FU (22.5 μM) inhibited 52% growth of the cells. All nine combinations of 6-gingerol with Ptx portrayed strong synergism with CI values of less than 0.4 [30]. This study displays the synergistic interactions of ginger phytochemicals when used in combination with each other. This shows the value of using whole extracts in traditional medicine and therefore the popularity of herbal medication unlike the use of synthetic drugs that are based on one active ingredient.

Synergism has been observed when crude water ginger extract was combined with Gelam honey and tested against colorectal cancer cell line HT29 [31]. The 3-(4,5-dimethylthiazol-2-ly)-2,5-diphenyltetrazolium bromide (MTT) assay of these combinations depicted synergism with CI_{50} values of less than 1 [32]. When tested singly against human colorectal adenocarcinoma cell line HT29, ginger extracts showed anticancer activity with an IC_{50} value of 5.2 mg/mL and Gelam honey showed an IC_{50} value of 80 mg/mL while a combination of an IC_{50} value of 0.3 mg/mL was observed [32]. These results indicate the activity of ginger is more effective when used in combination than when used alone against colorectal cancer [31, 32]. In a similar study, a combination of ginger extracts with Gelam honey showed synergistic activities (CI <1) with higher apoptotic activities when each is combined with 5-FU and tested against HCT 116 colon cancer cell lines [9].

Combinations of ginger extracts, especially combination of ginger phytochemicals, have possible use in treatment of cancer and reduction of emergence of anticancer drug resistance. With the increased emergence of anticancer drug resistance and the side effects of anticancer drugs, the use of safe and effective anticancer ginger combinations that do not have side effects is indispensable.

2.5 Antidiabetic activities of ginger combinations

Extracts of ginger, green tea (*Camellia sinensis* L.), and cinnamon (*Cinnamomum verum* J.Presl) were tested separately and then combined against diabetes and postprandial glucose patterns. The three herbs, ginger, green tea, and cinnamon, exerted a glycemic index (GI) of 72, 79, and 63, respectively, when tested singly using healthy human volunteers of both genders. Synergism was depicted when all extracts were combined with a GI of 60 [33]. Therefore, the highest effect of lowering postprandial glucose in humans was portrayed when the herbs were used in combination showing the potential use of ginger, green tea, and cinnamon combinations in controlling blood sugar level.

In another study by Ali et al. [34], aqueous ginger extracts showed anti-hyperglycemia and anti-inflammatory activities when tested alone against type 2 diabetes in rats, *in vivo* [35]. When the same aqueous extracts were combined with probiotics of eight good bacteria, they portrayed synergism. The combination showed substantial improvement in oral glucose tolerance test, serum insulin, C-peptide, and lipid abnormalities than when the species extracts were used alone [35].

Extracts of three commonly used culinary natural spices, ginger, garlic (*Allium sativum* L.), and cayenne pepper (*C. fructensces*) have shown prophylactic and therapeutic activities against cardiovascular diseases, atherosclerosis, and antidiabetic diseases [26, 36]. Synergism was observed when the three spices were tested in combination (1:1:1:1) against hypercholesterolemic activity.

The various phytochemicals in extracts of the spices acting in a beneficial manner explain the observed anti-hypercholesterolemic activity that was higher in combination through synergism than when the three spices were tested singly [26]. The use of ginger combinations in blood sugar regulation in traditional medicine is, therefore, preferred to the synthetic antidiabetic drugs, which rely on one single active ingredient. These studies depict a possible use of ginger combinations in blood sugar balance.

2.6 Antidepressant activities of ginger combinations

Depression is mainly caused by an imbalance of neurotransmitters; serotonin, norepinephrine, and dopamine. Antidepressants are used to maintain a balance of neurotransmitters, especially serotonin. Honokiol and magnolol (HMM) from magnolia bark (*Magnoliae officinalis* Rehd. Et Wils), and ginger essential oils (GEO) and polysaccharides (PGR) from both plant species were combined and tested for antidepressant activities in mice by measuring serotonin and noradrenaline levels in the hippocampus, prefrontal cortex, and striatum [37]. Most combinations of the compounds from the two plant species showed synergism. The most effective antidepressant synergistic effects were observed when 39 mg/kg of OGR were combined with 15 mg/kg of HMM after two weeks of antidepressant treatment. This resulted in a significant increase in serotonin and noradrenaline in the prefrontal cortex [37]. These results are in agreement with similar synergistic antidepressant effects of OGR in combination with HMM as reported by Qiang et al. [38].

Anxiety and depression are increased by mercury II chloride ($HgCl_2$) in mice [39]. In a study by Benkermiche et al. [40], ginger extracts combined with black cumin (*Nigella sativa* L.) oil showed prophylaxis synergistic effects on anxiety and depression behavior in Wistar rats that had been exposed to $HgCl_2$ [34]. This could be attributed to the pure compounds like 6-gingerol, isolated from ginger, which has shown neuroprotective effects in rodents [41], and the antidepressant activities of thymoquinone, the active ingredient in *N. sativa* which act synergistically when combined [14]. With increased reports of anxiety, depression, and even suicide, these studies are crucial as they are suggestive of possible use of ginger combinations both as a prophylaxis and therapeutic against depression.

2.7 Pharmacological activities of ginger combinations for herbal tea formulations

In production of herbal teas, various parts of medicinal plants are used; roots, stems, leaves, barks, flowers, and seeds. Most of these herbal teas are usually used in combination, packed in bags, and sold for use in different countries [42]. The herbal tea formulations are preferred because use of medicinal plants in combination leads to presence of various bioactive phytochemicals in one package, which has shown to reverse and prevent various metabolic diseases [40]. Studies of tea and herbal infusion combinations have shown increased pharmacological activities against various diseases [38]. Apart from the medicinal benefits, the use of these herbal tea formulations is promoted because consumers believe that they are natural and safe without any side effects [43]. In addition, herbal formulations, especially with ginger, provide tea with the desired taste and aroma [4].

In a study by Yusuf et al. [44], herbal tea formulations of powdered dried leaves of mango tree (*Mangifera indica* L.), leaves of drumstick tree (*Moringa oleifera* Lam.), and ginger powder have shown to have potential nutritional and health benefits in prevention of various metabolic diseases. Presence of bioactive phytochemical compounds such as tannins, flavonoids, terpenoids, and cardiac glycosides in these formulations indicate that such a composite blend is a reservoir of various antioxidants acting together in a beneficial way through synergism when used in combination thereby promoting health due to the therapeutic and healing properties of the phytochemicals contained in these plants [45]. In addition to the medicinal properties of ginger when used as a herbal tea formulation in combination with leaves of mango and drumstick tree, the combination provides herbal teas with the desired and appealing sensory attributes such as color, aroma, flavor, texture, taste, mouth feel, and overall acceptability [40]. This makes the medicinal use of ginger herbal tea formulations when combined with other natural spices preferable than standard approved drugs.

Ginger is used in coffee (*Coffea arabica* L.) herbal tea drinks. When used as an oxidant, ginger offers protection to the human body from various diseases that are attributed to reactions of radicals [7]. In this protection, ginger acts as a radical-scavenger, by inhibiting peroxidation and other free radical-mediated processes [13, 43]. When used alone, water extracts of ginger and coffee showed antiradical activities with EC_{50} values of 3.82 mg/mL and 1.78 mg/mL. Chlorogenic acid a pure compound isolated from ginger showed antiradical activity with an EC_{50} value of 36.76 µg/mL and caffeic acid isolated from coffee showed antiradical activity with an EC_{50} value of 22.37 µg/mL [13]. When coffee and ginger were combined, synergism was observed by the antiradical scavengers from extracts of the two plant species. This explains the medicinal benefits of the use of ginger when used in combination with herbal tea formulations for prevention of diseases.

When the plant extracts are digested, which imitates the drinking of herbal teas, the synergism increased *in vitro*. The combination of the pure compounds, chlorogenic acid from ginger and caffeic acid from coffee, showed antagonism for antiradical activity [13]. Phenolic compounds in ginger and coffee that are used in herbal teas act as an anti-inflammation through inhibition of pro-oxidative enzymes such as lipoxygenase (LOX) mediated arachidonic acid metabolism [20]. Water extracts of ginger and coffee showed LOX-inhibition activities with EC_{50} values of 3.83 mg/mL and 4.85 mg/mL, respectively. The combination of these extracts showed synergism as the Lox-inhibition activities of the pure compounds increased when digested *in vitro* [20]. Pure compounds, chlorogenic acid from ginger and caffeic acid from coffee, had a LOX-inhibition activity with EC_{50} values of 41.05 µg/mL and 22.86 µg/mL, respectively. When these compounds were combined, they showed stronger synergism [13]. Therefore, the ability of ginger to scavenge free radicals and inhibit lipoxygenase (LOX) is increased when ginger is used in combination with coffee. Since tea is the second most popular drink, second to water [12], the use of ginger in tea is, therefore, highly recommended as it makes the drink medicinal while improving the taste and aroma of tea.

2.8 Other pharmacological activities of ginger combinations

Synergism was shown when ginger was combined with nifedipine (a drug used to treat high blood pressure) in hypersensitive and normal patients. A combination of 1 g of ginger and 10 mg of nifedipine showed that a percentage inhibition of platelet induced by collagen, adenosine diphosphate (ADP), and epinephrine was 64.2%, 63.8%, 61.1%, respectively [44]. Ginger potentiated the antiplatelet aggression activity of nifedipine. The inhibition of platelet aggression suggests the use of ginger and nifedipine combination against cardiovascular and cerebrovascular complications [44].

Cisplatin is an anticancer drug in which reactive oxygen molecules (ROM) are involved in nephrotoxicity. Combinations of ginger ethanol extract (250 mg/kg of body weight) with vitamin E (α-tocopherol) (250 mg/kg) showed synergism against cisplatin-induced renal failure in mice [46]. In this combination, the activities of renal antioxidant enzymes such as superoxide dismutase (SOD), catalase (CAT), glutathione peroxidase (GPx), and glutathione (GSH) were increased while the level of malondialdehyde (MDA) was reduced [46]. The activity of the combination against cisplatin-induced acute renal failure is by increasing the antioxidant body's defense system. Therefore, the ginger-cisplatin combination aids the body's natural defense mechanism against renal failure. This study showed a potential use of ginger in combination with vitamin E against cisplatin-induced renal failure.

A specific ginger and glucosamine combination (Zinaxin Glucosamine; 170 mg EV.EXT 35 mixture and 500 mg glucosamine, as glucosamine sulfate, per capsule) showed potential use for mucosa protection by increased synthesis of mucosa prostaglandins (E1, E2, F2α, and 6-keto PGF1α) in knee and hip of osteoarthritis patients [47]. The gastrointestinal pain and dyspepsia were greatly reduced when compared with diclofenac. The study showed that this combination is as effective as diclofenac but has a better potential use as it concurrently provides gastro-protection, pain relief, and efficacy in osteoarthritis patients [47]. These extra health benefits in the use of natural remedies in phytomedicine make herbal medication popular, unlike pharmaceutical drugs that have adverse side effects. These results are in agreement with another recent study by Rondanelli et al. [48] on the use of ginger to relieve pain in osteoarthritis patients [49].

A study by Mustafa et al. [50] has shown that a combination of ginger powder and zinc supplements has potential pharmacological use in human health. This combination showed activity against oxidative damage, inflammation, and autophagy induced by fructose in rats with metabolic syndrome (MS). The combination also showed potential use in controlling glucose and lipid metabolism and the zinc homeostasis in rats with MS. The ginger powder and zinc supplement combination was shown to downregulate the expression of NF-κB, SREBP-1c, and mTORC1c genes and upregulate the expression of Nrf-2 and PPAR-α in the liver of rats with MS [51].

In another study, nanoparticles NP_S-PEG-FA loaded with the active ginger phytochemical 6-shogaol, demonstrated activity by alleviating colitis symptoms and increased colitis wound healing using mice model [30]. These results are suggestive of the therapeutic use of this approach in treatment of inflammatory bowels. In a similar study, semisolid poly (vinyl alcohol) hydrogels of ginger essential oil (GEO) encapsulation with chitosan nanoparticles (CNPs) displayed activity in wound healing. Increasing the amount of GEO in the hydrogels leads to a decrease in percentage encapsulation efficiency and increased percentage loading capacity [52, 53].

These studies on combination of ginger with standard approved drugs and nanoparticles reveal potential use of the medicinal plant species combinations in prevention and treatment of diseases. The use of these natural herbs combinations is also likely to reduce the emergence of resistance to these diseases.

3. Conclusion

In battling against various diseases and emergence of drug resistance, the use of two drugs in combination that have different modes of action greatly reduces the likelihood of the emergence of a drug-resistant strain. The higher the synergy the less the amount of each drug is required because less than 50% of each constituent should be able to achieve 100% treatment rates when synergy is found in a given combination [54]. Therefore, doses of drugs will be lower and thereby leading to much better dosing regimens, high tolerability, and safety [19, 55].

Synergism is distinctive in phytotherapy with the use of ginger in traditional medicine being efficacious and safe with no side effects [17]. Therefore, as epidemiological studies continue to report increased spread of various diseases and emergence of drug resistance, these factors argue for use of herbal-herbal combinations and herbal-drug combinations. In such combinations, numerous phytochemicals act in a useful manner and the activity is enhanced through synergism contrary to the use of synthetic pharmaceuticals that rely on one active ingredient [13]. This explains the increased popularity of herbal medication, especially when used in combination as is the case with phytochemical activities of ginger combinations.

Most of the pharmacological activities of ginger combinations have shown synergism conferred by its phytochemical components. The highest synergism was observed when phytochemicals isolated from ginger, gingerol, and shogaol derivatives were combined and tested against prostate cancer cell lines with an IC_{50} value of 0.03. Such synergistic interactions are vital in phytomedicine in overcoming the challenges of isolating active ingredients in medicinal plants since whole or partially purified extracts show efficacy in low doses [13]. These studies depict that ginger potentiates standard drugs, pure compounds, or other plant extracts when used in combination. Therefore, *in vivo* studies of ginger combinations and the mode of action are highly recommended as most of the ginger combination studies did not report on these aspects.

Author details

Douglas Ongeri Ochora
Department of Plant Sciences, Microbiology and Biotechnology, College of Natural Sciences, Makerere University, Kampala, Uganda

*Address all correspondence to: ochoraongeri@gmail.com

References

[1] Yuan H, Ma Q, Ye L, Piao G. The traditional medicine and modern medicine from natural products. Molecules. 2016;**21**:1-18

[2] Novotna B, Polesny Z, Pinto-Basto MF, Van Damme P, Pudil P, Mazancova J, et al. Medicinal plants used by 'root doctors', local traditional healers in Bié province, Angola. Journal of Ethnopharmacology. 2020;**19**:1-17

[3] Erhirhie EO, Ikegbune C, Okeke AI, Onwuzuligbo CC, Madubuogwu NU, Chukwudulue UM, et al. Antimalarial herbal drugs: A review of their interactions with conventional antimalarial drugs. Clinical Phytoscience. 2021;**7**:1-10

[4] Williamson EM. Synergy and other interactions in phytomedicines. Phytomedicine. 2001;**8**:401-409

[5] Singletary K. Ginger: An overview of health benefits. Nutrition Today. 2010;**45**:171-183

[6] Shahrajabian MH, Sun W, Cheng Q. Clinical aspects and health benefits of ginger (*Zingiber officinale*) in both traditional Chinese medicine and modern industry. Acta Agriculturae Scandinavica Section B: Soil and Plant Science. 2019;**69**:546-556

[7] Abbasi H, Khatoon R, Kabir H. *Zingiber officinale*: A simple spice with health benefits & some modern researches. Tang Humanitas Medicine. 2019;**9**:1-15

[8] Feng T, Su J, Ding ZH, Zheng YT, Li Y, Leng Y, et al. Chemical constituents and their bioactivities of "tongling White Ginger" (*Zingiber officinale*). Journal of Agricultural and Food Chemistry. 2011;**59**:11690-11695

[9] Gamage K, Dissanayake C, Angoda W, Chandrasiri WL, Liyanage RP. A review on medicinal uses of *Zingiber officinale* (Ginger). International Journal of Health Sciences and Research. 2022;**10**:142-148

[10] Munda S, Dutta S, Haldar S, Lal M. Chemical analysis and therapeutic uses of ginger (*Zingiber officinale* Rosc.) essential oil: A review. Journal of Essential Oil-Bearing Plants. 2018;**21**:994-1002

[11] Ghafoor K, Juhaimi F, Özcan MM, Uslu N, Babiker EE, Mohamed IA. Total phenolics, total carotenoids, individual phenolics and antioxidant activity of ginger (*Zingiber officinale*) rhizome as affected by drying methods. Food Science and Technology. 2020;**126**:1-7

[12] Mahboubi M. *Zingiber officinale* Rosc. essential oil, a review on its composition and bioactivity. Clinical Phytoscience. 2019;**5**:1-12

[13] Hakim L, Alias E, Makpol S, Ngah WZ, Morad NA, Yusof YM. Gelam honey and ginger Potentiate the anti cancer effect of 5-FU against HCT 116 colorectal cancer cells. Asian Pacific Journal of Cancer Prevention. 2014;**15**:4651-4657

[14] Yi LT, Xu Q, Li YC, Yang L, Kong LD. Antidepressant-like synergism of extracts from magnolia bark and ginger rhizome alone and in combination in mice. Progress in Neuro-Psychopharmacology & Biological Psychiatry. 2009;**33**:616-624

[15] Marx WM, Teleni L, Mccarthy AL, Vitetta L, Mckavanagh D, Thomson D, et al. Ginger (*Zingiber officinale*) and chemotherapy-induced nausea and vomiting: A systematic literature review. Nutrition Reviews. 2013;**71**:245-254

[16] Makanjuola S, Enujiugha V, Omoba O, Sanni D. Application of RSM and multivariate statistics in predicting antioxidant property of ethanolic extracts of tea-ginger blend. European Journal of Medicinal Plants. 2015;**6**:200-211

[17] Durak A, Gawlik-Dziki U, Kowlska I. Coffee with ginger – Interactions of biologically active phytochemicals in the model system. Food Chemistry. 2015;**166**:261-269

[18] Brahmbhatt M, Gundala SR, Asif G, Shamsi SA, Aneja R. Ginger phytochemicals exhibit synergy to inhibit prostate cancer cell proliferation. Nutrition and Cancer. 2013;**65**:263-272

[19] Lewandowska U, Gorlach S, Owczarek K, Hrabec E, Szewczyk K. Synergistic interactions between anticancer chemotherapeutics and phenolic compounds and anticancer synergy between polyphenols. Postępy Higieny i Medycyny Doświadczalnej. 2014;**68**:528-540

[20] Otunola GA, Afolayan AJ. Evaluation of the polyphenolic contents and some antioxidant properties of aqueous extracts of Garlic, Ginger, Cayenne Pepper and their mixture. Journal of Applied Botany and Food Quality. 2013;**86**:66-70

[21] Briskin DP. Medicinal plants and phytomedicines. Linking plant biochemistry and physiology to human health. Plant Physiology. 2000;**124**:507-514

[22] Ochora DO, Kakudidi EK, Namukobe J, Ipulet P, Wakoli DM, Okore W, et al. Synergism in antiplasmodial activities of artemether and lumefantrine in combination with *Securidaca longipedunculata*. Plants. 2022;**11**:1-14

[23] Gawlik-Dziki U. Dietary spices as a natural effectors of lipoxygenase, xanthine oxidase, peroxidase and antioxidant agents. Food Science and Technology. 2012;**47**:138-146

[24] Agarwal D, Sharma M, Dixit SK, Dutta RK, Singh AK, Gupta RD, et al. *In vitro* synergistic effect of fluoroquinolone analogues in combination with artemisinin against *Plasmodium falciparum*; Their antiplasmodial action in rodent malaria model. Malaria Journal. 2015;**14**:1-8

[25] Mathai K, Anand S, Aravind A, Dinatius P, Krishnan AV, Mathai M. Antimicrobial effect of ginger, garlic, honey, and lemon extracts on *Streptococcus mutans*. The Journal of Contemporary Dental Practice. 2017;**18**:1004-1008

[26] Al-terehi M, Al-saad AH, Al-musawi H, Zaida H. Synergism effect of antibacterial activity of some medicinal plants. Journal of Biology and Medical Sciences. 2013;**1**:46-53

[27] Betoni JC, Mantovani RP, Barbos LN, Di-Stasi LC, Fernandes A. Synergism between plant extract and antimicrobial drugs used on Staphylococcus aureus diseases. Memórias do Instituto Oswaldo Cruz. 2006;**101**:387-390

[28] Sebiomo A, Awofodu D, Awosanya O, Awotona F, Ajayi J. Comparative studies of antibacterial effect of some antibiotics and ginger (*Zingiber officinale*) on two pathogenic bacteria. Journal of Microbiology and Antimicrobials. 2011;**3**:18-22

[29] Maizura M, Aminah A, Aida WW. Total phenolic content and antioxidant activity of kesum (*Polygonum minus*), ginger (*Zingiber officinale*) and turmeric (*Curcuma longa*) extract. International Food Research Journal. 2011;**18**:526-531

[30] Otunola GA, Oloyede OB, Oladiji AT, Afolayan AJ. Selected spices and their combination modulate hypercholesterolemia-induced oxidative stress in experimental rats. Biological Research. 2014;**47**:2-7

[31] Putri DK, Fauzia S, Sabila I. Antioxidant activity from the combination ethanol extract secang wood (*Caesalpinia sappan* L .) and red ginger rhizome (*Zingiber officinale* Roxb.). Biological Research. 2021;**209**:143-147

[32] Zayadi RA, Abu-Bakar F, Ahmad MK. Elucidation of synergistic effect of *Eucalyptus globulus* honey and *Zingiber officinale* in the synthesis of colloidal biogenic gold nanoparticles with antioxidant and catalytic properties. Sustainable Chemistry and Pharmacy. 2019;**13**:1-8

[33] Mahomoodally MF, Aumeeruddy MZ, Rengasamy KR, Roshan S, Hammad S, Pandohee J, et al. Ginger and its active compounds in cancer therapy: From folk uses to nano-therapeutic applications. Seminars in Cancer Biology. 2021;**69**:140-149

[34] Ali HA, Mohamed SH, Alharbi HF, Algheshairy RM. Synergism between probiotics and herbs to manage Type 2 diabetes in rats. International Journal of Pharmacy and Pharmaceutical Sciences. 2020;**12**:26-35

[35] Zhang F, Zhang JG, Qu J, Zhang Q, Prasad C, Wei ZJ. Assessment of anti-cancerous potential of 6-gingerol (Tongling White Ginger) and its synergy with drugs on human cervical adenocarcinoma cells. Food and Chemical Toxicology. 2017;**109**:910-922

[36] Wee LH, Morad NA, Aan GJ, Makpol S, Wan N. Mechanism of chemoprevention against colon cancer cells using combined Gelam honey and Ginger extract via mTOR and Wnt/β-catenin pathways. Asian Pacific Journal of Cancer Prevention. 2016;**16**:6549-6556

[37] Tahir AA, Sani NA, Murad NA, Makpol S, Ngah WW, Yusof YM. Combined ginger extract & Gelam honey modulate Ras/ERK and PI3K/AKT pathway genes in colon cancer HT29 cells. Nutrition Journal. 2015;**14**:1-10

[38] Qiang LQ, Wang CP, Wang FM, Pan Y, Yi LT, Zhang X, et al. Combined administration of the mixture of honokiol and magnolol and ginger oil evokes antidepressant-like synergism in rats. Archives of Pharmacal Research. 2009;**32**:1281-1292

[39] Azzeh FS. Synergistic effect of green tea, cinnamon and ginger combination on enhancing postprandial blood glucose. In Pakistan Journal of Biological Sciences. 2013;**16**:74-79

[40] Benkermiche S, Djemli S, Haloui M, Bena-bed ML, Tahraoui A. Preventive effects of ginger extract and *Nigella sativa* oil on anxiety and depression behavior in wistar rats exposed to mercuric chloride. Pharmacognosy Research. 2021;**14**:1-4

[41] Owen PL, Johns T. Antioxidants in medicines and spices as cardioprotective agents in Tibetan highlanders. Pharmaceutical Biology. 2002;**40**:346-357

[42] Abu-Taweel GM, Rudayni HA. Curcumin ameliorated the mercuric chloride induced depression and anxiety in female mice offspring. Environmental Research. 2022;**204**:1-8

[43] Aquib M, Najmi AK, Akhtar M. Antidepressant effect of thymoquinone in animal models of depression. Drug Research. 2015;**65**:490-494

[44] Yusuf AB, Turaki AA, Adetunji AA. Formulation and evaluation of mango

leaf tea supplemented postprandial with moringa and ginger powder. Haya: The Saudi Journal of Life Sciences. 2022;7:151-157

[45] De-Heer B, Dakappa SS, Adhikari R, Timilsina SS, Sajjekhan SA. A review on the medicinal plant *Psidium guajava* linn. (Myrtaceae). Journal of Drug Delivery and Therapeutics. 2013;**3**:162-168

[46] Atoui AK, Mansouri A, Boskou G, Kefalas P. Tea and herbal infusions: Their antioxidant activity and phenolic profile. Food Chemistry. 2005;**89**:27-36

[47] Akila B, Vijayalakshmi R, Hemalatha G, Arunkumar R. Development and evaluation of property of guava leaf based herbal tea. Journal of Pharmacognosy functional and Phytochemistry. 2018;**7**:3036-3039

[48] Rondanelli M, Fossari F, Vecchio V, Gasparri C, Peroni G, Spadaccini D, et al. Clinical trials on pain lowering effect of ginger: A narrative review. Phytotherapy Research. 2020;**34**:2843-2856

[49] Okoli CO, Akah PA, Ezugworie U. Anti-Inflammatory activity of extracts of root bark of *Securidaca longipedunculata* Fres (Polygalaceae) African. Journal of Traditional, Complementary and Alternative Medicines. 2006;**3**:54-63

[50] Mustafa AZ, Hamed AR, Rostum AD, Abdulkarimi R, Abdulkareem NK, Akbari A. The combination of ginger powder and zinc supplement improves the fructose-induced metabolic syndrome in rats by modulating the hepatic expression of NF-κB, mTORC1, PPAR-α SREBP-1c, and Nrf2. Journal of Food Biochemistry. 2021;**45**:1-16

[51] Young HY, Liao JC, Chang YS, Luo YL, Lu MC, Peng WH. Synergistic effect of ginger and nifedipine on human platelet aggregation: A study in hypertensive patients and normal volunteers. The American Journal of Chinese Medicine. 2006;**34**:545-551

[52] Ajith TA, Nivitha V, Usha S. *Zingiber officinale* Roscoe alone and in combination with α-tocopherol protect the kidney against cisplatin-induced acute renal failure. Food and Chemical Toxicology. 2007;**45**:921-927

[53] Ngampunwetchakul L, Toonkaew S, Supaphol P, Suwantong O. Semi-solid poly(vinyl alcohol) hydrogels containing ginger essential oil encapsulated in chitosan nanoparticles for use in wound management. Journal of Polymer Research. 2019;**26**:224-232

[54] Ohrt C, Willingmyre GD, Lee P, Knirsch C, Milhous W. Assessment of azithromycin in combination with other antimalarial drugs against *Plasmodium falciparum In Vitro*. Antimicrobial Agents and Chemotherapy. 2002;**46**:2518-2524

[55] Drozdov VN, Kim VA, Tkachenko EV, Varvanina GG. Influence of a specific ginger combination on gastropathy conditions in patients with osteoarthritis of the knee or hip. Journal of Alternative and Complementary Medicine. 2012;**18**:583-588

Chapter 5

Golden Spice Turmeric and Its Health Benefits

Latika Yadav and Upasana

Abstract

Turmeric is a traditional spice extracted from the rhizomes of *Curcuma longa*, a ginger family member (Zingiberaceae). Turmeric, also known as the "Golden Spice of India," has been utilized for pharmacological purposes in India for ages. It has been used as a household remedy for biliary disorders, anorexia, cough, diabetic sores, hepatic disorders, rheumatism, and sinusitis in traditional medicine. Turmeric and its compounds, namely curcumin and essential oils, have a wide range of biological effects in addition to their usage as a spice and pigment. Curcumin, Turmeric's active ingredient, is being studied by scientists for its antioxidant activity, anti-inflammatory properties, anti-metabolic syndrome activities, neuroprotective activity, antimicrobial effects, anti-arthritis effects, anti-asthma, anti-obesity, cardio and liver toxicity protection activity, anti-depression and anxiety activities, anti-carcinogenic, anti-mutagenic, anticoagulant, anti-fertility activity, anti-diabetic, anti-fibrotic, anti-venom, anti-ulcer, hypotensive and hypocholesterolemic activities. As a result, turmeric and its compounds have the potential to be used in modern medicine to cure a wide range of diseases. These metabolic roles and actions of curcumin are depicted in this chapter for the benefit of human health.

Keywords: curcumin, turmeric, spice, antioxidant, health benefits

1. Introduction

The term "Turmeric" was derived from the Latin phrase "terra merita," which translates as "meritorious earth." Turmeric has been utilized in folk medicine and religious tradition for at least 6000 years and is also referred to as the "yellow root," the "golden spice," and "Indian saffron." It has 55 aliases in Sanskrit, each referring to a different part of religion or medicine (**Table 1**) [1]. Turmeric, the grounded rhizome of *Curcuma longa*, is a popular spice in curries and mustards. It frequently contributes to their distinctive color and flavor due to oleoresins and essential oils. Turmeric belongs to the ginger family (Zingiberaceae) illustrated in **Figure 1** is frequently used in traditional Chinese and Indian medicine to cure several ailments [2]. According to the Agricultural Market Intelligent Centre (PJTSAU), global turmeric production is approximately 11 lakh tonnes per year. India leads the global production scenario, accounting for 80% of the total, with China accounting for 8%, Myanmar accounting for 4%, Nigeria accounting for 3%, and Bangladesh accounting for 3%. India exported

Countries	Common Name
India	हल्दी *(haldi) in Hindi;* हरिद्रा *(haridrā)* or वरवर्णिनी *(varavarṇinī)* in Sanskrit
Nepal	हल्दी *(haldi)*and *(be-sar)*
Bangladesh	Haldi, halud
Pakistan	Haldi
Japan	欝金 *(ukon)*, Tamerikku
China	姜黄 (*jiang huang*, literal: "Ginger Yellow")
United States	Indian saffron,turmeric
Germany	*Gelbwurz* (literal: Yellow Root) or *Kurkuma*
Russia	Koren, kurkumy, Kurkuma
Portugal	Açafrão da Índia, Curcuma
Korea	(강황)*kang hwang*
Thailand	ขมิ้น *(kamin)*, Khamin,khamin-chan
Malayasia	Kunyit, temu kuniyit
Spain	Curcuma, Azafran arabe
France	Curcuma, Safran des Indes
Indonesia	Koneng,kunir,kuniyit,tius
Arabia	كركم *(kurkum)*
Persia	زردچوبه *(zardchubeh)*

Source: https://www.newworldencyclopedia.org/entry/Turmeric.

Table 1.
Common names of C. longa *in different countries.*

1.71 lakh tonnes of turmeric in 20–21, up from 1.37 lakh tonnes the previous year. Bangladesh (49,522 tonnes), the UAE (12,182 tonnes), Iran (10,964 tonnes), the United States (9,712 tonnes), and Morocco are the top turmeric importing countries from India (8,522 tonnes). Turmeric sales have been steadily increasing since the COVID-19 outbreak in 2020–2021 [3].

C. longa L. grown in hot and humid climates. It needs much water and grows up to one-meter height. Turmeric plant leaves are enormous and oblong with short pseudostem. It carries pale yellow flowers without fruits. The rhizome grown underground contains a mother rhizome with many branching subordinate rhizomes. They are ovate, oblong, or pyriform in shape and pale yellow, reddish-yellow, or orange-brown [4–6] shown in **Figure 2** and **Table 1**.

2. Phytoconstituents of turmeric

Turmeric has been found to have over 100 constituents. Turmeric's primary root component is a volatile oil containing turmerone and additional colorants called curcuminoids. Turmeric's principal phytoconstituents are diarylheptanoids, which combine to form curcuminoids and account for approximately 16% of turmeric's dry weight [7]. The majority of crude turmeric extracts, as well as some refined

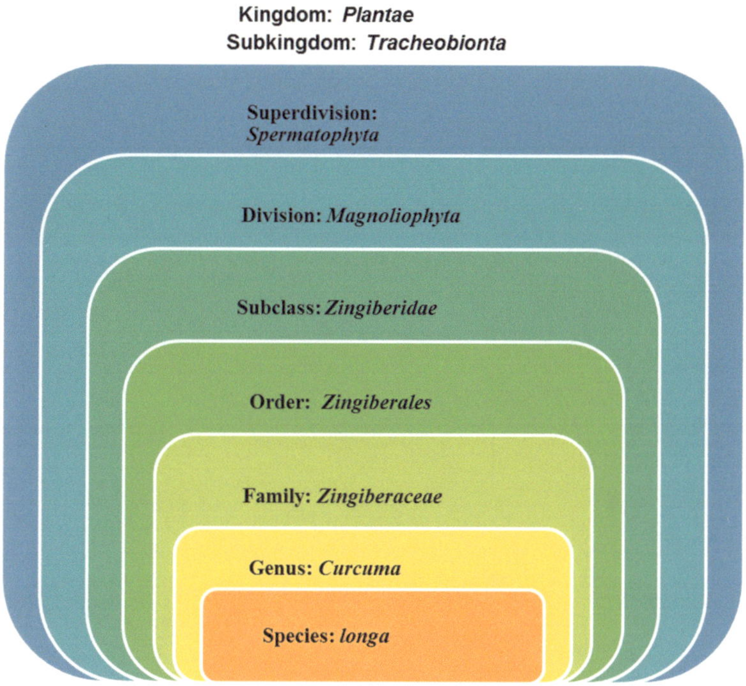

Figure 1.
Venn diagram shows the taxonomical classification of Curcuma longa.

"curcumin" materials (**Figure 3**), contain three major compounds: curcumin I (diferuloylmethane) at 94%, curcumin II (demethoxycurcumin) at 6%, and curcumin III (bisdemethoxycurcumin) at 3%, in addition to volatile oils, sugars, proteins, and resins [8]. Turmeric in its purest form comprises 5–6.6 percent curcumin, 0.5 percent extraneous matter, 3% mold, and 3.5 percent volatile oils. Turmerone, arturmerone, curcumene, germacrone, and ar-curcumene are examples of these compounds [9]. Volatile oils include d-α-phellandrene, d-sabinene, cinol, borneol, zingiberene, and sesquiterpenes [10]. Turmerone, arturmerone, and zingiberene are the active ingredients in turmeric that give it its flavor and aroma. Additionally, four novel polysaccharides, ukonans, stigmasterole, β-sitosterole, cholesterol, and 2-hydroxymethyl anthraquinone, were discovered in the rhizomes in recent investigations [11, 12].

According to the Indian Food Composition Table (2017), Turmeric Powder (*Curcuma domestica*) is a major source of macro and micronutrients. The **Table 2** shows that it is rich in fibers, vitamins, and minerals. Turmeric is also a good source of linoleic acid and α-linolenic acid [13].

According to several studies, curcumin is "generally recognised as safe" (GRAS) as a food additive up to a dose of 20 mg per serving, according to the FDA [14]. Together with turmeric's long history and cultural use as a medicine, this classification has contributed to its appeal as a dietary supplement marketed for a range of common ailments. Curcumin supplement sales were estimated to have exceeded $20 million in the United States in 2014, but an exact figure is difficult to verify [15]. The Dietary Supplement Health and Education Act legislation constructing the validity of dietary supplements in the United States (1994) and progressions in *in-vitro* testing almost certainly played a significant role in a dramatic increase in the publication of manuscripts describing the use of curcumin in biological studies in the late 1990s

Figure 2.
Pictures of Curcuma Longa *plant (A), flower (B), turmeric rhizomes (C) and rhizome powder (D) of turmeric* (Curcuma longa *L.). The images used in drawing the figure were extracted from the following links as described below: (A) https://www.amazon.in/Creative-Farmer-Turmeric-Medicinal-Ayurvedic/dp/B08P5S58V1, (B) https://ellegadodenewton.com/2020/06/10/es-la-curcuma-el-alimento-ayurveda-milagroso-el-caso-de-la-actividad-biologica-de-la-curcuma/, (C) https://dimsaleglobal.com/product/fresh-tumeric/, and (D) https://www.benessereblog.it/cose-il-latte-doro-e-come-si-prepara.*

(**Figure 4**). Curcumin has been shown to have anti-inflammatory, anti-cardiovascular, antibacterial, antifungal, antiviral, antidiabetic, skin protective, radioprotective, wound healing, antigastrointesinal properties, antioxidant, immunomodulating, anticarcinogenic, and Alzheimer's [16].

3. Health benefits of curcumin

Turmeric's key element, curcumin, has been demonstrated to have a various health benefits.

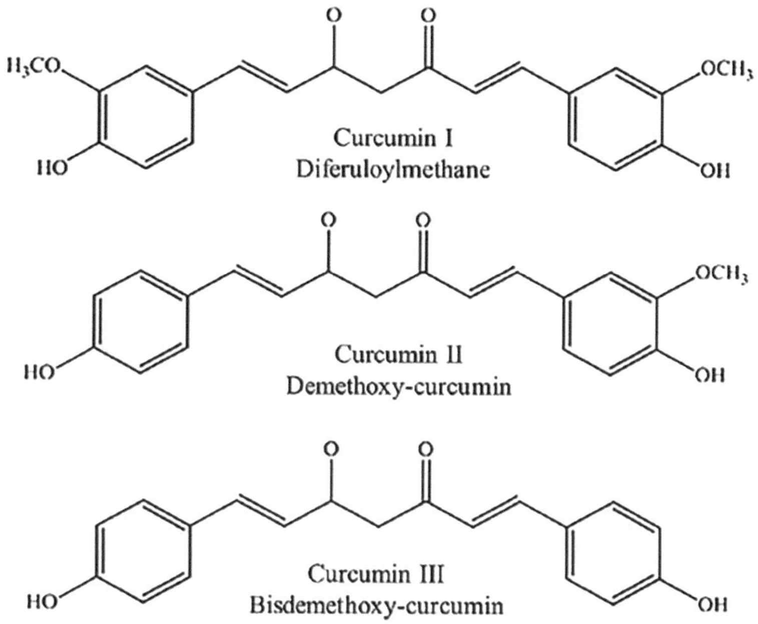

Figure 3.
Structure of Curcumin I (Diferuloylmethane), Curcumin II (Demethoxy-curcumin) and Curcumin III (Bisdemethoxy-curcumin). Source: https://www.tumbral.com/tag/desmethoxycurcumin.

Constituents	Composition
Moisture	10.58 g/100 g
Protein	7.66 g/100 g
Ash	6.13 g/100 g
Total Fat	5.03 g/100 g
Total Dietary Fiber (Insoluble and Soluble)	21.38 (18.79 and 2.59)g/100 g
Carbohydrate	49.22 g/100 g
Energy	1174KJ
Thiamine B1	0.06 mg/100 g
Riboflavin B2	0.01 mg/100 g
Niacin B3	1.55 mg/100 g
Pantothenic acid B5	0.13 mg/100 g
Total B6	0.13 μg/100 g
Biotin B7	0.76 μg/100 g
Total folate B9	13.86 μg/100 g
Linoleic Acid	1563 mg/100 g
α-Linolenic acid	377 mg/100 g
Total Saturated Fatty Acid (TSFA)	1634 mg/100 g
Total Mono Unsaturated Fatty Acid (TMUFA)	448 mg/100 g
Total Poly Unsaturated Fatty Acid (TPUFA)	1940 mg/100 g

Constituents	Composition
Lutein	99.8 μg/100 g
Zeaxanthin	3.56 μg/100 g
Beta Carotene	55.20 μg/100 g
Total Carotenoid	427 μg/100 g
Calcium	122 mg/100 g
Iron	46.08 mg/100 g
Copper	0.44 mg/100 g
Magnesium	260 mg/100 g
Sodium	24.41 mg/100 g
Potassium	2374 mg/100 g
Phosphorus	276 mg/100 g
Zinc	2.64 mg/100 g
Selenium	6.41 μg/100 g

Table 2.
*Nutrient composition of turmeric powder (*Curcuma domestica*).*

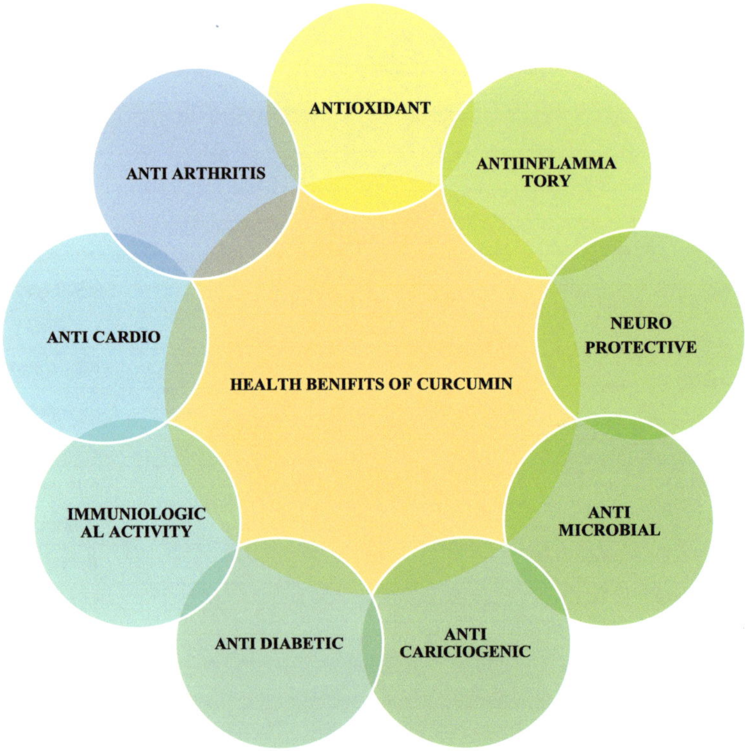

Figure 4.
Health benefits of curcumin.

3.1 Antioxidant activity

Curcumin has been shown to protect against oxidative damage during indomethacin-induced gastric lesions by inhibiting gastric peroxidase inactivation and directly scavenging H_2O_2 and OH. Since reactive oxygen species have been associated with the development of various pathological illnesses, Turmeric's potent antioxidant activity enables it to regulate these diseases [17]. In vivo, curcumin reduces the formation of reactive oxygen species. It was reported that by suppressing lipid peroxidation and by scavenging a plethora of reactive oxygen species such as hydrogen peroxide, Nitric oxide radicals, and superoxide radicals, curcumin enhances its antioxidant functions. This latter action is related to the increased activity of numerous antioxidant enzymes, including Superoxide dismutase, Catalase and Glutathione peroxidase [18]. It was also reported that curcumin boosts GSH levels by increasing the expression of glutathione transferase and its mRNAs, regarded as a chain-terminating antioxidant due to its lipophilic properties and decreased reactive oxygen species production by enzymes such as lipoxygenases, cyclooxygenases, and xanthine oxidase [19].

3.2 Role in anti-inflammation

Inflammation is a vital activity in the body because it helps the body fight off invading microorganisms and repair damage caused by bacteria, viruses, and traumas. Numerous studies have demonstrated that curcumin has significant promise for treating various inflammatory illnesses [20–22]. Curcumin is a potent anti-inflammatory agent that inhibits both lipoxygenase and COX-2. Both in vitro and in vivo investigations have established its anti-inflammatory properties in the acute and chronic phases of inflammation. Curcumin reduced edema in mice at dosages ranging from 50 to 200 mg/kg. A 48 mg/kg body weight dose resulted in a 50% reduction in edema, making curcumin approximately as efficacious as cortisone and phenylbutazone at comparable doses. A lower dose of 20–80 mg/kg reduced paw inflammation and edema in rats. Curcumin also reduced formaldehyde-induced arthritis in rats when administered at a 40 mg/kg dose and exhibited no acute toxicity when administered at levels up to 2 g/kg/day [23].

3.3 Help in treating rheumatoid arthritis and osteoarthritis

RA is a chronic, progressive autoimmune disease characterized by severe and symmetric polyarthritis. Numerous studies have demonstrated that Curcumin possesses anti-arthritic effects. Curcumin and rapamycin dramatically reduced ankle and joint redness and swelling in rheumatoid arthritis rats. Curcumin blocked the mTOR pathway generated by CIA and the invasion of inflammatory cells into the synovium induced by RA. Curcumin and rapamycin therapy decreased proinflammatory cytokine levels in CIA rats, including IL-1, TNF-, MMP-1, and MMP-3 [24]. Additionally, it was discovered that consuming turmeric extracts alone or in combination with other herbal substances can help control pain and improve function in persons with knee osteoarthritis [25].

3.4 Prevent from cardiovascular disease

Dyslipidemia is a significant and prevalent risk factor for cardiovascular disease in the general population. Turmeric may be helpful in preventing arterial blockage, which can result in either a heart attack or a stroke. Turmeric contributes to maintaining normal cholesterol levels and inhibits LDL cholesterol oxidation (bad cholesterol). Oxidized LDL deposits in artery walls and contributes to atherosclerotic plaque development. Additionally, turmeric may inhibit platelet aggregation along injured blood vessel walls. Platelets deposit together at the site of a ruptured blood vessel, leading to the formation of blood clots and arterial obstruction [26]. Curcumin has been shown to improve endothelial function and decrease monocyte adhesion generated by TNFα in endothelial cells via NF-κB suppression [27]. Additionally, it was shown that curcumin inhibits the production of the angiotensin II type 1 receptor, hence reducing cardiovascular disorders. Curcumin inhibits the AT1R gene promoter's ability to bind to the specificity protein 1 [28]. Turmeric and curcumin were shown to protect people at risk of cardiovascular disease by improving serum lipid levels [29].

3.5 Effect on gastrointestinal tract

Turmeric possesses several anti-inflammatory characteristics that are beneficial to the digestive tract. The components of *Curcuma longa*, sodium curcuminate, and p-tolymethylcarbinol have various positive effects on the digestive tract. Due to curcumin's increased bioavailability in the intestinal tract, gastrointestinal illnesses such as inflammatory bowel disease, hepatic fibrosis, and gastrointestinal malignancies have been among the most investigated ailments, demonstrating curcumin's potential therapeutic benefit [30]. Sodium curcuminate decreases intestinal spasm and the release of p-tolymethylcarbinol while boosting the secretion of gastrin, secretin, bicarbonate, and pancreatic enzymes. Curcumin was demonstrated to protect the mucosa of mice with artificially induced colitis from harm. Curcumin was able to reduce inflammation in experimentally induced pancreatitis rats significantly. Curcumin was also reported to inhibit the production of pro-inflammatory mediators in other kinds of induced pancreatitis, such as cerulean or ethanol, as determined by histology, pancreatic trypsin, serum amylase, and neutrophil infiltration [31]. Turmeric has also been demonstrated to prevent ulcer formation in rats exposed to various gastrointestinal stimuli, including stress, alcohol, indomethacin, pyloric ligation, and reserpine [32].

3.6 Antidiabetic properties

Turmeric has been shown in experimental studies to play a substantial effect on diabetes. Turmeric rhizome powder is beneficial in Madhumeha (diabetes mellitus) when combined with Amla juice and honey [33]. Turmeric's active components, curcuminoids, inhibit lipid peroxidation by promoting the activity of antioxidant enzymes such as superoxide dismutase, catalase, and peroxidase. Curcumin and its three derivatives (demethoxycurcumin, bisdemethoxycurcumin, and diacetyl curcumin) are responsible for *C. longa*'s antioxidant capabilities [34]. It has been shown that the ethanolic extract of turmeric, which contains curcuminoids and sesquiterpenoids, is significantly more hypoglycemic than curcuminoids or sesquiterpenoids alone. Turmeric has remarkable effects on postprandial plasma glucose and insulin levels [35, 36]. Turmeric also helps prevent problems associated with diabetes mellitus. Turmeric's impact on blood sugar was demonstrated experimentally on albino rats, and the polyol pathway discovered that both turmeric and curcumin lowered blood sugar levels in alloxan-induced diabetes [37].

3.7 Effect on skin

Due to its antioxidant properties, it has photoprotective properties. Unsaturated lipids make up a sizable portion of the lipids on the skin's surface. As a result, they are frequently targeted by free radicals. The sun's UV rays penetrate the epidermis, accelerating the oxidative damage produced by free radicals. Prolonged exposure to these radiations may damage the lipids, resulting in a loss of skin texture. Turmeric extract has been proven in laboratory experiments to reduce inflammation and protect epidermal cells from the damage caused by ultraviolet B radiation. Curcumin has been proven to protect against chromosomal damage caused by gamma radiation in tiny dosages of turmeric [38].

3.8 Anti-cancer effect

Curcumin has been found to inhibit carcinogenesis via two distinct mechanisms: angiogenesis and cancer cell proliferation. Additionally, it inhibits cancer cell metastasis and promotes apoptosis in cancer cells [39]. Curcumin has been demonstrated to inhibit angiogenic factor stimulators such as VEGF and primary fibroblast growth factor. Indeed, curcumin has been shown to inhibit VEGF expression via NF-kB and AP-1 regulation, inhibiting IL-8 expression [40]. Curcumin is also capable of inducing apoptosis in cancer cells via a p53-dependent mechanism. p53 is a well-characterized tumor suppressor protein that regulates cell proliferation, necrosis, and DNA damage [41]. Curcumin has also been shown to reduce colon cancer by suppressing the Wnt/b-catenin signaling pathways via miR-130a [42]. In cancer models, curcumin has also been shown to inhibit and suppress the PI3K/Akt signaling pathway [43, 44].

3.9 Immunological activity

Curcumin demonstrates a broad range of biological activities that benefit human health. Apart from these functions, curcumin's primary property is immunological activity, which is why it has been demonstrated to be effective against anti-immune diseases. When the effect of curcumin on the immunological profiles of the blood was explored, it was shown that white blood cells (WBCs) and especially lymphocytes produced increased levels of immunoglobulins (IgG and IgM) [45]. Curcumin has been shown to attenuate bradykinin-induced coughing [46, 47]. It inhibits chemokine release and, therefore, may protect against acute lung injury [46, 48]. Patients with severe respiratory sickness exhibit a hyper-immune response manifested by extensive alveolar destruction, epithelial apoptosis, fibrin accumulation, and the creation of a hyaline membrane. A low neutrophil to lymphocyte ratio is a poor prognostic predictor for COVID-19 at the cellular and molecular level [49, 50].

3.10 Antimicrobial properties

The rhizome of *C. longa* has traditionally been employed as an antibacterial agent [51]. Numerous studies have demonstrated curcumin's antimicrobial activity, which includes antibacterial, antiviral, and antifungal properties:

3.10.1 Antibacterial activity

Antibacterial activity against Staph. Epidermis ATCC 12228, Staph. aureus ATCC 25923, *Klebsiella pneumoniae* ATCC 10031, and *E. coli* ATCC 25922 were demonstrated using an aqueous extract of *C. longa* rhizome minimum inhibitory concentration values of 4–16 g/L and minimum bactericidal concentration values of 16–32 g/L [52]. Likewise, it was demonstrated that adding 0.3 percent (w/v) aqueous curcumin extract to cheese reduced *Salmonella typhimurium*, *Pseudomonas aeruginosa*, and *E. coli* 0157:H7 bacterial counts. Additionally, it reduced *Staph. Aureus, Bacillus cereus*, and *Listeria monocytogenes* contamination following a 14-day cold storage period [53]. Turmeric oil was also efficient against *B. subtilis*, *B. coagulans*, *B. cereus*, Staph. aureus, *E. coli*, and *P. aeruginosa* as a byproduct of curcumin synthesis [54].

3.10.2 Antiviral activity

Curcumin, a plant product, has been shown to exhibit a broad spectrum of antiviral action against various viruses. Due to its rate-limiting action in the de novo synthesis of guanine nucleotides, the inosine monophosphate dehydrogenase (IMPDH) enzyme has been proposed as a therapeutic target for antiviral and anticancer agents. Among the 15 specific polyphenols, curcumin has been recommended as a powerful antiviral agent via its inhibitory activity against IMPDH in either a non-competitive or competitive way [55]. Additionally, curcumin dramatically suppressed the acetylation of the HIV Tat protein by p300, which is related to the inhibition of HIV-1 replication. Curcumin is a potent chemical for combinatorial HIV treatments because it targets the p300/CREB-binding protein (CBP) acetyltransferase proteins [56].

3.10.3 Antifungal activity

Historically, extracts from various natural resources, mainly plants, served as an effective armament for battling fungal infections and rotting. Curcumin inhibited the growth of two phytophagous fungi, *Fusarium solani and Helminthosporium oryzae*. Turmeric oil was highly effective against *F. solani* and *H. oryzae*, with an IC50 of 19.73 and 12.7 g/mL, respectively [57]. Turmeric powder was applied at concentrations of 0.8 and 1.0 g/L to plant tissue culture and showed a substantial inhibitory effect against fungal contaminations [58]. Curcumin also inhibited *Cryptococcus neoformans* and *Cryptococcus dubliniensis*, with a minimum inhibitory concentration (MIC) of 32 mg/L [59]. Curcumin was discovered to be a highly effective fungicide against 14 Candida strains, including four ATCC strains and ten clinical isolates, with MIC values ranging from 250 to 2000 g/mL [60].

4. Conclusions

Medicinal plants are a veritable goldmine of materials containing active compounds. They play a crucial part in both traditional and novel medication development. The importance of medicinal plants in human health, cultural values, and well-being is recognized in many parts of the world. Nowadays, the demand for herbal medications is approximately 80% of the population and increases as people become more reliant on herbal remedies. A comprehensive review of the literature found that *C. longa*, with its various pharmacological properties, is regarded as a comprehensive global approach among herbal medicines. Curcumin is used as a culinary spice and food color and an ingredient in various Ayurvedic and Chinese medical formulations. Through the decades, science has established the broad range of beneficial impacts curcumin has on human health. In India, the "golden spice" has been used for centuries as a cooking ingredient, preservative, coloring agent for dye, traditional medicines, ceremonies, and religious purposes, as well as a home treatment. In India, every household uses turmeric daily. This plant is regarded as a versatile medicinal plant because it contains various chemical components. Thus, it is evident that in order to battle diseases, extensive research is required to determine their therapeutic utility.

Acknowledgements

We pay our profound sense of gratitude to Dr. Satish Kumar Yadav for his assistance, encouragement, and insightful advice throughout in constructing this book chapter. We also apologize for not citing the research papers of all the authors that helped me in better understanding this topic.

Conflict of interest

Authors declare no conflict of interest.

Author details

Latika Yadav[1*] and Upasana[2]

1 Government Degree College, Punwarka, Saharanpur, Uttar Pradesh, India

2 Mahila College Dalmianagar, Rohtas, Veer Kunwar Singh University, Ara, Bihar, India

*Address all correspondence to: drlatika27@gmail.com

References

[1] Ravindran PN, Nirmal Babu K, Sivaraman K. Turmeric: The goldenspice of life. In: Turmeric: The Genus Curcuma. Boca Raton: CRC Press; 2007. pp. 1-14

[2] Kuttan R, Bhanumathy P, Nirmala K, George MC. Potential anticancer activity of turmeric (*Curcuma longa*). Cancer Letters. 1985;**29**:197-202

[3] Agricultural Market Intelligent Centre (PJTSAU) Turmeric outlook, January. 2022. Available from: https://pjtsau.edu.in/agri-marketing-intelligence.html [Accessed: January 1, 2022]

[4] Parthasarathy V, Chempakam B, Zachariah T. Chemistry of Spices. Wallingford, UK: CABI Pub; 2008

[5] Trujillo J, Chirino YI, Molina-Jijón E, Andérica-Romero AC, Tapia E, Pedraza-Chaverrí J. Renoprotective effect of the antioxidant curcumin: Recent findings. Redox Biology. 2013;**1**:448-456

[6] WHO. Monographs on Selected Medicinal Plants-Volume 1: Rhizoma Curcumae Longae. Apps. who.int. 2016. Available from: http://apps.who.int/iris/bitstream/handle/10665/42052/9241545178.pdf [Accessed: January 1, 2022]

[7] Niranjan A, Singh S, Dhiman M, Tewari SK. Biochemical composition of *Curcuma longal*. Accessions Analytical Letters. 2013;**46**:1069-1083

[8] Nasri H, Sahinfard N, Rafieian M, Rafieian S, Shirzad M, Rafieian-kopaei M. Turmeric: A spice with multifunctional medicinal properties. Journal of HerbMed Pharmacology. 2014;**3**:5-8

[9] Chanda S, Ramachandra TV. Phytochemical and pharmacological importance of turmeric (*Curcuma longa*): A review. Research & Reviews: A Journal of Pharmacology. 2019;**9**:16-23

[10] Ohshiro M, Kuroyanag M, Keno A. Structures of sesquiterpenes from *Curcuma longa*. Phytochemistry. 1990; **29**:2201-2205

[11] Kapoor LD. Handbook of Ayurvedic Medicinal Plants. Boca Raton, FL: CRC Press; 1990

[12] Kirtikar KR, Basu BD, Blatter E, Caius JF, Mhaskar KS. Indian Medicinal Plants. 2nd ed. Allahabad, India: Lalit Mohan Basu; 1993. p. 1182

[13] Longvah T, Ananthan R, Bhaskarachary K, Venkaiah K. 2017. National Institute of Nutrition. Indian Council of Medical Research, Department of Health Research, Ministry of Health and Family Welfare, Government of India

[14] Food and Drug Administration Office of Food Additive Safety. Agency Response Letter GRAS Notice No. Grn 000460. U.S. Food and Drug Administration. 2013

[15] Majeed S. The state of the curcumin market. Natural Products Insider; Informa Exhibitions. 2015. Available from: http://www.naturalproductsinsider.com/articles/2015/12/the-state-of-the-curcumin-market.aspx [Accessed: January 6, 2022]

[16] Nelson KM, Dahlin JL, Bisson J, Graham J, Pauli GF, Walters MA. The essential medicinal chemistry of Curcumin. Journal of Medical Chemistry. 2017;**60**:1620-1637

[17] Uday B, Dipak D, Banerjee Ranajit K. Reactive oxygen species: Oxidative

damage and pathogenesis. Current Science. 1999;77:658-666

[18] Ak T, Gulcin I. Antioxidant and radical scavenging properties of curcumin. Chemico Biological Interactions. 2008;**174**:27-37. DOI: 10.1016/j.cbi.2008.05.003

[19] Priyadarsini KI, Maity DK, Naik GH, Kumar MS, Unnikrishnan MK, Satav JG, et al. Role of phenolic O–H and methylene hydrogen on the free radical reactions and antioxidant activity of curcumin. Free Radical Biology and Medicine. 2003;**35**: 475-484. DOI: 10.1016/s0891-5849(03) 00325-3

[20] Cianciulli A, Calvello R, Porro C, Trotta T, Salvatore R, Panaro MA. PI3k/Akt signalling pathway plays a crucial role in the anti-inflammatory effects of curcumin in LPS-activated microglia. International Immunopharmacology. 2016;**36**:282-290. DOI: 10.1016/j.intimp.2016.05.007

[21] Dai W, Wang H, Fang J, Zhu Y, Zhou J, Wang X, et al. Curcumin provides neuroprotection in model of traumatic brain injury via the Nrf2-ARE signaling pathway. Brain Research Bulletin. 2018;**140**:65-71. DOI: 10.1016/j.brainresbull.2018.03.020

[22] Edwards RL, Luis PB, Varuzza PV, Joseph AI, Presley SH, Chaturvedi R, et al. The anti-inflammatory activity of curcumin is mediated by its oxidative metabolites. Journal of Biological Chemistry. 2017;**292**:21243-21252. DOI: 10.1074/jbc.RA117.000123

[23] Rao CV, Rivenson A, Simi B, Reddy BS. Chemoprevention of colon carcinogenesis by dietary curcumin, a naturally occurring plant phenolic compound. Cancer Research. 1995;**55**: 259-266

[24] Dai Q, Zhou D, Xu L, Song X. Curcumin alleviates rheumatoid arthritis-induced inflammation and synovial hyperplasia by targeting mTOR pathway in rats. Drug Design, Development and Therapy. 2018;**2018**(12):4095-4105. DOI: 10.2147/DDDT.S175763

[25] Natural Medicines Comprehensive Database. 2020. Turmeric: Uses, Side Effects, and More. Available from: https://www.webmd.com/vitamins/ai/ingredientmono-662/turmeric. [Accessed: January 06, 2022]

[26] Salehi B, Del Prado-Audelo ML, Cortés H, Leyva-Gómez G, Stojanović-Radić Z, Singh YD, et al. Therapeutic applications of curcumin nanomedicine formulations in cardiovascular diseases. Journal of Clinical Medicine. 2020;**9**:746. DOI: 10.3390/jcm9030746

[27] Monfoulet LE, Mercier S, Bayle D, Tamaian R, Barber-Chamoux N, Morand C, et al. Curcumin modulates endothelial permeability and monocyte transendothelial migration by affecting endothelial cell dynamics. Free Radical Biological Medicine. 2017;**112**:109-120. DOI: 10.1016/j.freeradbiomed.2017.07.019

[28] Yao Y, Wang W, Li M, Ren H, Chen C, Wang J, et al. Curcumin exerts its anti-hypertensive effect by down-regulating the AT1 receptor in vascular smooth muscle cells. Scientific Reports. 2016;**6**:25579. DOI: 10.1038/srep25579

[29] Qin S, Huang L, Gong J, Shen S, Huang J, Ren H, et al. Efficacy and safety of turmeric and curcumin in lowering blood lipid levels in patients with cardiovascular risk factors: A meta-analysis of randomized controlled trials. Nutrition Journal. 2017;**16**:68. DOI: 10.1186/s12937-017-0293-y

[30] Rajasekaran SA. Therapeutic potential of curcumin in gastrointestinal

diseases. World Journal of Gastrointestinal Pathophysiology. 2011; **2**:1-14. DOI: 10.4291/wjgp.v2.i1.1

[31] Velayudhan KC, Dikshit N, Nizar MA. Ethnobotany of turmeric (*Curcuma longa* L.). Indian Journal of Traditional Knowledge. 2012;**11**:607-614

[32] Pawar MA, Patil SS, Nagrik DM. Phytochemical and physicochemical investigation of *Curcuma longa* Linn rhizome. International Journal of Chemical and Physical Sciences. 2015;**4**: 458-463

[33] Acharya YT. Charaka Samhitha of Agnivesh with the Ayurveda Dipika Commentary. 4th ed. Varanasi, India: Chaukambha Sanskrit Samstha; 1994. p. 447

[34] Faizal IP, Suresh S, Satheesh Kumar R, Augusti KT. A study on the hypoglycemic and hypolipidemic effects of an ayurvedic drug rajanyamalakadi in diabetic patients. Indian Journal of Clinical Biochemistry. 2009;**24**:82-87

[35] Ammon HP, Wahl MA. Pharmacology of *Curcuma longa*. Planta Medicine. 1991;**57**:1-7

[36] Saxena J, Sahu R. Evaluation of phytochemical constituents in conventional and non-conventional species of Curcuma. International Research Journal of Pharmacy. 2012;**3**: 203-204

[37] Shrishail D, Harish KH, Ravichandra HG. Tulsian and, Shruthi S. Turmeric: Nature's precious medicine Asian Journal of Pharmaceutical and Clinical Research. 2013;**6**:10-16

[38] Khajehdehi P. Turmeric: Reemerging of a neglected Asian traditional remedy. Journal of Nephropathology. 2012;**1**: 17-22

[39] Sharifi-Rad J, Rayess YE, Rizk AA, Sadaka C, Zgheib R, Zam W, et al. Turmeric and its major compound Curcumin on Health: Bioactive effects and safety profiles for food, pharmaceutical, biotechnological and medicinal applications. Frontiers in Pharmacology. 2020;**11**:01021. DOI: 10.3389/fphar.2020.01021

[40] Yance DR, Sagar SM. Targeting angiogenesis with integrative cancer therapies. Integrative Cancer Therapies. 2006;**5**:9-29. DOI: 10.1177/1534735405285562

[41] Kandoth C, McLellan MD, Vandin F, Ye K, Niu B, Lu C, et al. Mutational landscape and significance across 12 major cancer types. Nature. 2013;**502**: 333-339. DOI: 10.1038/nature12634

[42] Dou H, Shen R, Tao J, Huang L, Shi H, Chen H, et al. Curcumin suppresses the colon cancer proliferation by inhibiting Wnt/β-Catenin pathways via miR-130a. Frontiers in Pharmacology. 2017;**8**:877. DOI: 10.3389/fphar.2017.00877

[43] Kasi PD, Tamilselvam R, Skalicka-Woźniak K, Nabavi SF, Daglia M, Bishayee A, et al. Molecular targets of curcumin for cancer therapy: An updated review. Tumour Biology. 2016; **37**:13017-13028. DOI: 10.1007/s13277-016-5183-y

[44] Rana C, Piplani H, Vaish V, Nehru B, Sanyal SN. Down regulation of PI3-K/Akt/PTEN pathway and activation of mitochondrial intrinsic apoptosis by Diclofenac and Curcumin in colon cancer. Molecular and Cellular Biochemistry. 2015;**402**:225-241. DOI: 10.1007/s11010-015-2330-5

[45] Nawab A, Tang S, Li G, An L, Wu J, Liu W, et al. Dietary curcumin supplementation effects on blood

immunological profile and liver enzymatic activity of laying hens after exposure to high temperature conditions. Journal of Thermal Biology. 2020;**90**:102573. DOI: 10.1016/j.jtherbio.2020.102573

[46] Momtazi-Borojeni AA, Haftcheshmeh SM, Esmaeili SA, Johnston TP, Babaei F, Nassiri-Asl M, et al. Curcumin (a constituent of turmeric): New treatment option against COVID-19. Food Science and Nutrition. 2020;**8**:5215-5227. DOI: 10.1002/fsn3.1858

[47] Ahmadi R, Salari S, Sharifi MD, Reihani H, Rostamiani MB, Behmadi M, et al. Oral nano-curcumin formulation efficacy in the management of mild to moderate outpatient COVID-19: A randomized triple-blind placebo-controlled clinical trial. Food Science and Nutrition. 2021;**9**:4068-4075. DOI: 10.1002/fsn3.2226

[48] Jyoti K, Pandey RS, Kush P, Kaushik D, Jain UK, Madan J. Inhalable bioresponsive chitosan microspheres of doxorubicin and soluble curcumin augmented drug delivery in lung cancer cells. International Journal of Biological Macromolecules. 2017;**98**:50-58. DOI: 10.1016/j.ijbiomac.2017.01.109

[49] Jimeno S, Ventura PS, Castellano JM, García-Adasme SI, Miranda M, Touza P, et al. Prognostic implications of neutrophil-lymphocyte ratio in COVID-19. European Journal of Clinical Investigation. 2021;**51**:e13404. DOI: 10.1111/eci.13404

[50] Ulloque-Badaracco JR, Ivan Salas-Tello W, Al-Kassab-Córdova A, Alarcón-Braga EA, Benites-Zapata VA, Maguiña JL, et al. Prognostic value of neutrophil-to-lymphocyte ratio in COVID-19 patients: A systematic review and meta-analysis. International Journal of Clinical Practice. 2021;**75**:e14596. DOI: 10.1111/ijcp.14596

[51] Rudrappa T, Baish HP. Curcumin, a known phenolic from Curcuma longa, attenuates the virulence of *Pseudomonas aeruginosa* PAO1 in whole plant and animal pathogenicity models. Journal of Agricultural and Food Chemistry. 2008;**56**:1955-1962

[52] Niamsa N, Sittiwet C. Antimicrobial activity of *Curcuma longa* aqueous extract. Journal of Pharmacology and Toxicology. 2009;**4**:173-177

[53] Hosny MI, El KWI, Murad HA, El DRK. Antimicrobial activity of Curcumin upon pathogenic microorganisms during manufacture and storage of a novel style cheese 'Karishcum'. Journal of American Science. 2011;**7**:611-618

[54] Negi PS, Jayaprakasha GK, Rao LJM, Sakariah KK. Antibacterial activity of turmeric oil: A byproduct from curcumin manufacture. Journal of Agricultural and Food Chemistry. 1999;**47**:4297-4300

[55] Chen D-Y, Shien J-H, Tiley L, Chiou S-S, Wang S-Y, Chang T-J, et al. Curcumin inhibits influenza virus infection and haemagglutinaton activity. Food Chemistry. 2010;**119**:1346-1351

[56] Zandi K, Ramedani E, Mohammadi K, Tajbakhsh S, Deilami I, Rastian Z, et al. Evaluation of antiviral activities of curcumin derivatives against HSV-1 in Vero cell line. Natural Product Communications. 2010;**5**:1935-1938

[57] Balasubramanyam K, Varier RA, Altaf M, et al. Curcumin, a novel p300/CREB-binding protein-specific inhibitor of acetyltransferase, represses the acetylation of histone/nonhistone proteins and histone acetyltransferase-

dependent chromatin transcription. The Journal of Biological Chemistry. 2004; **279**(49):51163-51171

[58] Chowdhury H, Banerjee T, Walia S. In vitro screening of Curcuma longa L and its derivatives as antifungal agents against Helminthosporrum oryzae and Fusarium solani. Pesticide Research Journal. 2008;**20**:6-9

[59] Upendra RS, Khandelwal P, Reddy AHM. Turmeric powder (Curcuma longa Linn.) as an antifungal agent in plant tissue culture studies. International Journal of Engineering Science. 2011;**11**:7899-7904

[60] Martins CVB, da Silva DL, ATM N, Magalhães TFF, Watanabe GA, Modolo LV, et al. Curcumin as a promising antifungal of clinical interest. Journal of Antimicrobial Chemotherapy. 2009;**63**:337-339. DOI: 10.1093/jac/dkn488

Chapter 6

The Therapeutic and Phytopharmacological Potential of Ginger (*Zingiber officinale*)

Madonna Ngwatshipane Mashabela and Wilfred Otang-Mbeng

Abstract

Ginger (*Zingiber officinale*), a Zingiberacae family member, is a popular spice all around the world. This perennial creeping plant has long leaves, vivid green flowers, and a strong tuberous rhizome. Throughout history, several cultures and civilizations have recognized the potential of ginger in the treatment and prevention of disease. Various research with ginger and its extract support the potential effect in a variety of applications. Ginger has been used medicinally for over 2000 years and is regarded as one of the most versatile medicinal herbs, with a wide range of biological properties. Since ancient times, ginger has been used to cure a variety of conditions including heart problems, menstrual disorders, food poisoning, osteoarthritis, epilepsy, nausea, inflammation, cough and cold, motion sickness, menstrual cramps, cancer, and many others. Apart from that, it has antibacterial and antioxidant properties. The presence of gingerol and paradol, as well as shogaols and other chemicals, is responsible for ginger's medicinal properties. The medical benefits of ginger and current knowledge provide a solid platform for future research into how it can protect humans from a number of diseases.

Keywords: medicinal uses, gingerol, COVID-19, antioxidants, nausea

1. Introduction

Medicinal plants are the "backbone" of traditional medicine, which means people in the less developed countries utilize medicinal plants on a regular basis. These medicinal plants are thought to be a rich source of components for medication discovery and synthesis. Furthermore, these plants are important in the evolution of human cultures all around the world. Plants have formed the foundations of conventional traditional medical systems for thousands of years. Plants will continue to provide us with new remedies. Some of the beneficial benefits attributed to plants have been proven to be false, and medical plant therapy is based on hundreds to thousands of years of experimental research [1]. Many cultures around the world rely on traditional medicine since it is conveniently accessible from a local plant. According to a WHO (World Health Organization) estimate, the medicinal plant is used by over 3.5 billion people in underdeveloped nations for health care [2–4]. Traditional medicine is still prescribed by African doctors, i.e. (70–80%). Traditional medicine is used to treat human and livestock ailments in almost every part of the world [5]. A lot of attention has been paid to various areas of ginger's activities in reviews. For example, Grzanna et al. [6] focused on the use of ginger as an anti-inflammatory agent, whereas Shukla and Singh [7] focused on the cancer prevention properties of the crude drug. Chaiyakunapruk et al. [8] published a review on the effects of ginger as a post-operative anti-emetic drug. As a result, the goal of this chapter was to present an overview of the medicinal and phytopharmacological potential of ginger (*Zingiber officinale*).

2. Description of ginger

Z. officinale belongs to the Zingiberaceae plant family, which includes 1300 species and 49 genera, with 80–90 Zingiber species. It is native to East and Southern Asia. Its common name, Zingiber, derives from the Greek word zingiberis from the Sanskrit word singabera, which means "spice." The Latin word, Zingiber, means "horn-shaped," and refers to the roots' antler-like appearance. Ginger is a weedy perennial herbaceous plant that grows to a height of 2 meters and thrives in humid environment [9]. Some of the names given to ginger are Zenzero in Italian, Jeung or Sang Keong in Chinese, Aliah in Indonesia, Adrack in Urdu, Gember in Dutch, Jengibre in Spanish, Ingwar in German, and Gingembre in French [10]. Ginger was one of the first eastern spices introduced to Europe, and it is still in high demand today [10]. Zingiberaceae species typically have thickened rhizomes with secretory cells producing essential oil [11].

Ginger is traded in three basic forms - green (fresh), pickled or preserved and dry. Only dry ginger (whole, peeled, or sliced) is considered a spice; green or fresh ginger is primarily used as a vegetable, while pickled or preserved ginger is primarily used in the Chinese and Japanese culinary industries. In addition, ginger oil and oleoresins are also available for purchase. Despite the fact that ginger is grown in many countries, India and China are the top two exporters of dry ginger, followed by Nigeria, Sierra Leone, Australia, Fiji, Bangladesh, Jamaica, Nepal and Indonesia.

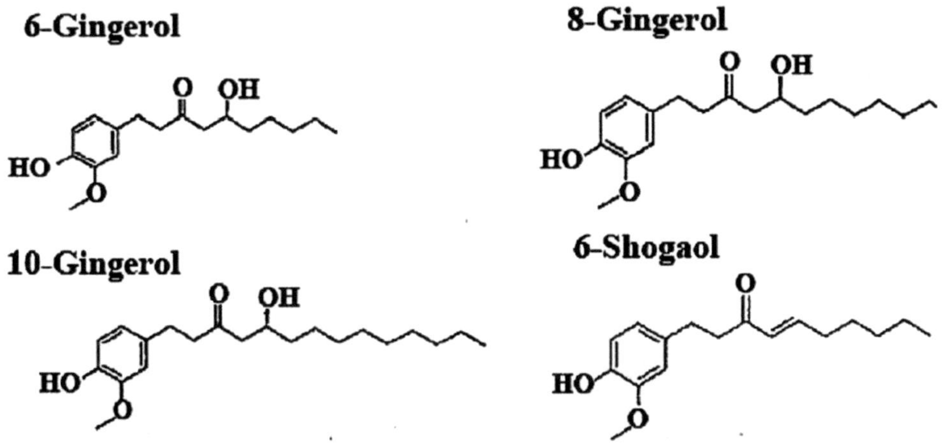

Figure 1.
Structure of 8-gingerol, 6-gingerol, 6-shogaol and 10-gingerol [14].

2.1 Ginger's active ingredients

Ginger has over 60 constituents, all of which are considered active compounds [12]. Alantolucton, inulin and certain essential oils are key components found in this plant root [12]. The powdered rhizome is made up of 9% protein, 3–8% crude fiber, 3–6% fatty oil, 60–70% carbohydrates, 9–12% water, roughly 8% ash and 2–3% volatile oil [13]. Gingerols, Ingenol, gingerdiols, zingerone, paradols, and shogaol are just a few of the therapeutic substances found in ginger. Gingerols are the most important of these chemicals because of their active pharmacological characteristics. Gingerols and shogaol are chemically composed of 6-, 8- and 10-structural analogs (**Figure 1**) [15].

3. Discussion

3.1 Therapeutic and Phytopharmacological potential of ginger

The most important therapeutic properties of *Z. officinale* (ginger) are given below:

3.1.1 Evidence of ginger's anti-COVID-19 properties

SARS-CoV-2-related papain-like protease (PLpro) cleaves polyprotein a/b (PP a/b) at different sites yielding several proteins needing for viral survival and replication (**Figure 2**) [17]. SARS-CoV-2-related PLpro also interferes with type I IFN anti-virus response [17]. In order to successfully suppress virus multiplication and survival, anti-SARS-CoV-2 medications should target PLpro. [18]. According to the findings of a study conducted in Saudi Arabia, COVID-19 patients' consumption of ginger increased from 36.2% prior to infection to 57.6% after infection. The proportion of patients admitted to the hospital for COVID 19 treatment was also lower (28.0%) among ginger users than among nonusers (38.0%) [19]. A few examples of cured COVID-19 patients were recorded in a Bangladesh study, who drank home medicines using ginger in blends of various herbs with or without further treatments [20]. A few cases of COVID-19 were treated with home remedies including ginger in combination with other herbs improved disease symptoms, according to the findings of a Tunisian study [21]. In certain African countries, renowned treatments involving ginger in combinations with other herbs were also utilized to treat COVID-19 [22]. In compared to those treated with a normal protocol using hydroxy-chloroquine alone, the results of an Iranian clinical trial study show that combining outpatients with probable COVID-19, ginger and Echinacea alleviated some of their clinical symptoms (breath shortness, coughing, and muscular discomfort) [23]. Ginger may help patients with pulmonary problems such ARDS (Acute respiratory distress syndrome), fibrosis, lung and pneumonia, as well as sepsis, which are all indications seen in COVID-19. Overall, the evidence suggests that more high-quality controlled trials are needed to validate ginger's benefit and safety in COVID-19 patients. In Iran, a clinical investigation is underway in which 84 COVID-19 patients were randomly assigned to two groups, each with 42 participants, including intervention and control groups [24]. The intervention group will receive standard treatment plus 1000 mg ginger three times daily for seven days, whereas the control group will receive normal treatment plus placebo tablets at the same dose and schedule for seven days [16].

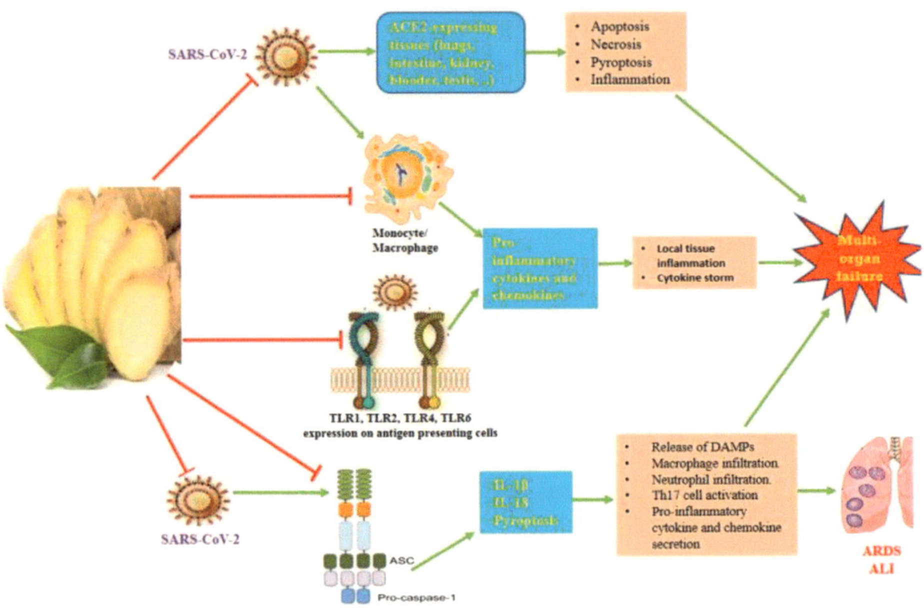

Figure 2.
COVID-19 pathogenesis is influenced by SARS-CoV-2 and inflammation. SARS-CoV-2 infects tissues that express ACE2, causing necrosis, pyroptosis and apoptosis, as well as inflammation. Infected macrophages with SARS-CoV-2 release a variety of cytokines and chemokines, causing tissue inflammation and a cytokine stormTLRs and inflammasomes can be activated by SARS-CoV-2, resulting in an increase in inflammatory responses and tissue damage. Ginger can enhance antiviral immune responses and have direct anti-SARS-CoV-2 effects, as well as interfere with inflammatory responses mediated by macrophages, TLRs, and inflammasomes. Abbreviations: ALI: Acute lung injury; ACE2: Angiotensin-converting enzyme 2; SARS-CoV-2: Severe acute respiratory syndrome coronavirus 2; ARDS: Acute respiratory distress syndrome; TLR: Toll-like receptor; IL: Interleukin [16].

4. Medicinal uses and potential health benefits in traditional medicine

Ginger has direct antimicrobial properties and can thus be used to treat bacterial infections [25]. It is used as a stimulant in Traditional Chinese Medicine to treat colic and atonic dyspepsia [26–28]. Ginger is a Yang plant that might help to reduce Yin and nourish the body [29]. Ginger is described as spicy and hot in Traditional Chinese Medicine, and it is said to treat cold extremities and warm up the body, enhance a slow and erratic heartbeat, a pale appearance on skin, and strengthen the body after blood loss [30]. Ginger is used as a herbal treatment for a variety of cardiovascular conditions [31]. Other researchers emphasized that ginger is used to treat inflammation, nausea, headaches, arthritis, muscular discomfort, rheumatism, and colds in Ayurvedic, Chinese, Arabic, and African traditional remedies (**Figure 3**) [32, 33]. The rhizomes of ginger have recently been utilized in Traditional Medicine to treat a variety of cardiovascular disorders, including hypertension [34]. Ginger has been used as an anti-edema drug in Iranian Traditional Medicine as a therapy for a variety of diseases, such as athero-sclerosis, gastric ulcer, respiratory disorders, gastrointestinal disorders, migraine, cholesterol; depression and nausea, other benefits of ginger include pain relief, anti-inflammatory, rheumatoid arthritis and antioxidant effects [35]. This is one of India's most popular spices, and it has long been used in traditional oriental medicine to treat common colds, stomach problems, and rheumatism [36]. The primary components of ginger include 10-gingerol, 8-gingerol, 6-shogaol, and

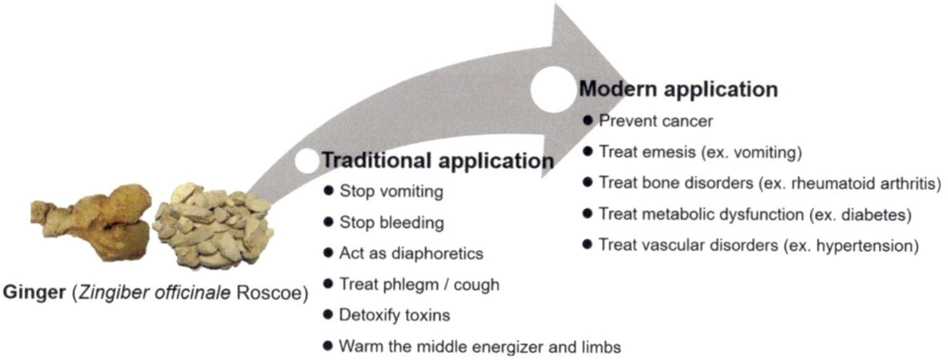

Figure 3.
Traditional and modern pharmacological applications of ginger. In both traditional and modern medicine, ginger has been used to treat a variety of symptoms and disorders [32].

Phenols	Volatile Sesquiterpenes	Others
Shogoals and Gingerols	Sesquiphellandrene, Bisapolene, Zingiberene, Zingiberol, Curcurmene	Zingerone, 6-dehydrogingerdione, Galanolactone, Gingesulfonic acid, Neral, Geraniol, Gingerglycolipids Monoacyldigalactosylglycerols,

Table 1.
Ginger's active chemical components [41].

6-gingerol, all of which have been demonstrated to have significant antioxidant activity [37]. Ginger's most prevalent bioactive ingredient, 6-gingerol, has a wide range of pharmacological actions, including antipyretic activities, antioxidant, anti-inflammatory and analgesic [38, 39]. After being digested and absorbed by the digestive system, shogaols can be partially changed to paradols by cooking or metabolized to paradols in the animals' bodies [40]. Anti-inflammatory and antioxidant activities are known to exist in Shogaol and Gingerol in particular (**Table 1**) (**Figure 1**) [42].

5. Medicinal uses and potential health benefits in modern medicine industry

As a powerful anti-oxidant, ginger extract can treat illnesses induced by oxidative stress. Extant anthocyanins and phenolic compounds, such as sugevals and gingerols, have been shown to have neuroprotective qualities, such as memory enhancement, analgesic effects, and learning induced by aging [43]. Ginger can be used in a variety of dishes, both sweet and savory, including drinks, puddings, apple pie, cakes, breads, sweets, and sauces, soups, savory puddings, grills, roasts, stews, and more [44]. Furthermore, heat had no influence on the active hypoglycaemic component of ginger, as a result, ginger consumption in both cooked and raw forms in a variety of cuisines could be an effective diabetes management strategy. Ginger contains anti nausea, antimicrobial, cardio tonic, antiemetic, antipyretic, anti-hypertensive, analgesic, anti ulcer, hypoglycaemic, anti-inflammatory, and anti-platelet aggregation effects in both experimental animals and human patients [41]. Antibacterial agents, gingerols, phenylbutenoids, diterpenoids, shogaols, diarylheptanoids, sesquiterpenoids, and flavanoids, as well as phenylbutenoids, diarylheptanoids, flavonoids, and diterpenoids, were discovered in phytochemical examination of many species of ginger rhizomes [45, 46]. Because ginger leaves have more antioxidant activity than rhizomes and flowers, it has been proven in some studies that it has significant potential for development into functional foods and other health products [47]. An infusion of ginger rhizomes with brown sugar is used to cure common colds, according to Semwal et al. [15], while scrambled eggs with powdered ginger is used as a home remedy in China to relieve coughing. Ginger is used in the United States to treat motion sickness and morning sickness during pregnancy, as well as leg cramps [15]. Maghbooli et al. [48] confirmed the effectiveness of ginger powder in the treatment of common migraine attacks and its antiepileptic drug-like properties. Ginger has been found in numerous trials to help prevent cancer and to treat nausea and vomiting caused by pregnancy and chemotherapy [49–51].

Ginger has the ability to preserve DNA from harm caused by H_2O_2 and may be useful in improving sperm quality [52]. Ginger rhizome has long been used in Iran to improve male sexuality, control female menstrual cycles, and relieve painful periods [53]. Mahassni and Bukhari [54] revealed that the extract of ginger rhizome had varied effects on immune system cells and antibodies in smokers and non-smokers, despite the fact that both benefited from thyroid gland augmentation. Furthermore, ginger may help smokers with anemia, whereas it may help non-smokers have a higher humoral immunity or antibody response to infections. According to Atashak et al. [55], both ginger supplementation and progressive resistance training (PRT)

Treats diarrhea
Relieves nausea
Prevents menstrual cramps
Prevents obesity
Treats cold and flu
Prevents infection
Improves cognition
Skin care
Aids in proper digestion
Reduces arthritis pain
Stomach ulcers
Detoxifies the body
Relives asthma
Relieves muscle pain
Liver protection
Prevents cancer
Controls diabetes
Boosts heart health
Increases sexual activity
Improves brain function
Removes excess gas
Regulates blood sugar Regulates blood sugar
Regulates blood sugar
Regulates blood sugar
Regulates blood sugar

Table 2.
The most significant advantages of ginger.

protect against oxidative stress after 10 weeks, therefore both therapies can help obese persons. Future trials focused on the efficacy of *Z. officinale* in treating various forms of pain, as well as treatment duration, dosing methodologies, safety, and cost efficiency, will be built on the foundation laid by Gagnier et al. [56]. **Table 2** lists the most major health advantages of ginger [41].

5.1 Blood pressure-lowering effect

Ginger has been used for a long time to treat a variety of ailments, including hypertension. The crude extract of ginger induced a dose-dependent (0.3–3 mg/kg) fall in the arterial blood pressure of anesthetized rats and a cardio depressant activity in guinea pigs. The blood-pressure-lowering effect of ginger is mediated through blockade of voltage-dependent calcium channels [57]. Another study found that aqueous ginger extract reduces blood pressure by a dual inhibitory effect mediated by muscarinic receptor stimulation and Ca^{++} channel blocking, and this work provides a solid mechanistic basis for the use of ginger in hypertension and palpitations [57]. The effect of ginger tea on blood pressure of hypertensive individuals has been studied to determine and compare the mean arterial pressure, median number of hypertensive episodes of the respondents with and without intake of ginger tea. From the fifth to the eighth week, when the participants began drinking 10 g of ginger tea twice a day, their average mean arterial pressure dropped to 94.804 mmHg [58].

5.2 Other medicinal and toxicological properties

Muscular aches, indigestion, fever, hypertension, fever, rheumatism, pains, infectious diseases, sore throats, cramps, colds, vomiting, constipation, motion sickness, gastrointestinal diseases, arthritis, sprains, helminthiasis and dementia are just a few of the ailments that ginger has been used to treat in Chinese, Ayurvedic, Tibb-Unani, Sri Lankan, Arabic, and African traditional medicines. Ginger has been traditionally used from time immemorial for varied human ailments in different parts of the world, to aid digestion and treat stomach upset, diarrhea, nausea, and migraine. In a double-blind, randomized research involving 60 ladies who have undergone major gynecological surgery the group that got ginger root had much less reported instances of nausea and was comparable to the placebo group [59]. In a study of 100 persons with common migraine headaches, Maghbooli et al. [48], compared the efficiency of ginger (250 mg of ginger rhizome powder) to sumatriptan (50 mg, a synthetic migraine headache medication) (no aura). Both medicines resulted in a 44% reduction in pain 2 hours after treatment, though it is unclear how much of this was due to natural history and/or a placebo effect. Both medicines resulted in a 44% reduction in pain 2 hours after treatment, though it is unclear how much of this was due to natural history and/or a placebo effect.

6. Ginger's memory-boosting properties

As the population ages, aging-induced cognitive impairment is recognized as a prodromal stage of dementia that affects the quality of life of older people [60, 61]. Then, in order to slow the course of dementia, researchers may need to look into interventions that improve memory function. Ginger improved memory function in aging-related disorders, according to a clinical trial by Saenghong et al. [62]. They tested middle-aged women's working memory and cognitive function after giving them ginger extract orally, and found that the ginger treatment increased scores in word recognition, digit vigilance, choice reaction, arithmetic working memory, and spatial working memory significantly. Ginger treatment improved memory in normal mice via increasing hippocampus levels of nerve growth factor (NGF), which activates extracellular signal–regulated kinases (ERK) and then cAMP response element–binding protein (CREB), resulting in increased synaptogenesis [63–65]. In these investigations, ginger was found to help with cognitive decline in the early stages of dementia in the elderly.

6.1 Effects of ginger and its compounds on stroke

Stroke incidence rates and bad outcomes both rise with age, indicating that more than 80% of strokes occur in those over the age of 65 [66]. After a stroke, brain damage is more severe in the elderly, and endogenous synthesis of antioxidant enzymes and neurotropic factors is lacking [67]. The loss of ovarian hormones with age, as well as the functional decline of brain supporting cells including astrocytes and endothelial cells, contribute to the severity of stroke outcomes in the aging brain [68]. Two separate studies looked into the pharmacological potential of ginger and found that it can help with memory loss and brain infarction caused by middle cerebral artery occlusion (MCAO) [29, 69].

6.2 Ginger's anti-epileptic properties

Ginger has been shown to have extraordinary effects on epilepsy patients who suffer from recurring seizures. When compared to younger individuals, epilepsy in the elderly has fewer pre-symptoms and a longer duration of postictal disorientation [70–72]. This effect could be linked to the substantial changes in brain anatomy that occur as a result of ventricular expansion speeding up as people get older [73]. Ginger extract was effective for reducing the duration and incidence of seizures induced by pentylenetetrazole (PTZ), which is a well-known inducer of epilepsy-like symptoms in an age-dependent manner [74, 75]. Another study conducted lately in mice found comparable benefits after treatment with ginger extract [76]. Ginger therapy significantly reduced myoclonic and clonic seizures, as well as forelimb tonic extension, in both investigations. These effects of ginger could be attributed to 6-gingerol's inhibitory effects on the synthesis of nitric oxide (NO), which activates the soluble guanylyl cyclase, which controls the seizure threshold [77].

7. Conclusion

Ginger is a popular condiment, cooking spice, and herbal remedy around the world, as well as a flavoring agent. Ginger contains over 60 active ingredients, which are classified as non-volatile or volatile. Hydrocarbons mostly monoterpenoid hydrocarbons and sesquiterpene include the volatile component of ginger and impart distinct aroma and taste to ginger. Paradols, gingerols, zingerone and shogaols, are examples of non-volatile chemicals. Ginger's active components, such as gingerols, shogaols, zingerone, and others, have antioxidant properties. 6-gingerol and 6-shogaol are the most pungent gingerols and shogaols found in the rhizome. The main ingredient in ginger, gingerol, has been extracted and tested for toxic and pharmacological effects. Cold-induced disease, colic, dyspepsia, loss of appetite, swellings, heart palpitation, nausea, asthma, cough and rheumatism have all been treated with fresh ginger. Anti-inflammatory qualities, anti-thrombotic properties, cholesterol-lowering characteristics, blood pressure-lowering properties, anti-microbial properties, anti-oxidant properties, anti-tumor properties, and hypoglycaemic activities are all medicinal properties linked with ginger. Ginger can help with cancer, obesity, heart disease, osteoarthritis, hypertension, bacterial infections and diabetes among other things. Ginger is a herbal, readily available, low-cost treatment with a low risk profile that can be used in place of chemical, rare, and expensive pharmaceuticals. Ginger can alter critical fundamental processes involved in COVID-19 formation because of its antiviral, anti-inflammatory, immunomodulatory, and antioxidant properties. This study summarizes current understanding about the possibility of ginger and its components for the treatment of COVID-19. Ginger appears to have some promising health advantages, according to other scientific literature, and further data from additional clinical research will help confirm whether ginger's many health benefits can be realized in humans. Herbal remedies and other nutraceuticals are increasingly and extensively used by a substantial part of the population. To sum up, treatment with natural herbal medicine especially ginger, non-synthetic drug, is recommended. However, the majority of these pharmacological benefits of ginger must be confirmed in clinical research before the pharmacological usefulness of ginger and its ingredients may be endorsed.

Author details

Madonna Ngwatshipane Mashabela[1,2*] and Wilfred Otang-Mbeng[1]

1 Faculty of Agriculture and Natural Sciences, School of Biology and Environmental Sciences, University of Mpumalanga, Mbombela, South Africa

2 Agricultural Research Council, Infruitec-Nietvoorbij, Private Bag, Stellenbosch, South Africa

*Address all correspondence to: wilfred.mbeng@ump.ac.za

References

[1] Beyene B, Beyene B, Deribe H. Review on application and management of medicinal plants for the livelihood of the local community. Journal of Resources Development and Management. 2016;**22**(1):33-39

[2] World Health Organization. WHO Medicines Strategy 2004-2007: Countries at the Core (No. WHO/EDM/2004.5). Geneva, Switzerland: World Health Organization; 2004

[3] Bekele G, Reddy PR. Ethnobotanical study of medicinal plants used to treat human ailments by Guji Oromo tribes in Abaya District, Borana, Oromia, Ethiopia. Universal Journal of Plant Science. 2015;**3**(1):1-8

[4] Seifu T, Asres K, Gebre-Mariam T. Ethnobotanical and ethnopharmaceutical studies on medicinal plants of Chifra district, Afar region, North Eastern Ethiopia. Ethiopian Pharmaceutical Journal. 2004;**24**(1):41-58

[5] Meragiaw M. Wild useful plants with emphasis on traditional use of medicinal and edible plants by the people of Aba'ala, North-Eastern Ethiopia. Journal of Medicinal Plant and Herbal Therapy Research. 2016;**4**(1):1-16

[6] Grzanna R, Phan P, Polotsky A, Lindmark L, Frondoza CG. Ginger extract inhibits beta-amyloid peptide-induced cytokine and chemokine expression in cultured THP-1 monocytes. Journal of Alternative and Complementary Medicine. 2004;**10**:1009-1013

[7] Shukla Y, Singh M. Cancer preventive properties of ginger: A brief review. Food and Chemical Toxicology. 2007;**45**:683-690

[8] Chaiyakunapruk N, Kitikannakorn N, Nathisuwan S, Leeprakobboon K, Leelasettagool C. The efficacy of ginger for the prevention of postoperative nausea and vomiting: A meta-analysis. American Journal of Obstetrics and Gynecology. 2006;**194**:95-99

[9] Teimoory H, Azizi M, Najafi MF, Behzadi A, Rezaei M. Antibacterial activity of Myrtus communis L. and Zingiber officinale rose extracts against some gram positive pathogens. Research Opinions in Animal and Veterinary Sciences. 2013;**3**(12):478-481

[10] Ravindran PN and Babu KN editos. Ginger: The genus Zingiber. Boca Raton, Florida, USA: CRC Press; 2016

[11] Aye YY. Microscopical Characters, Phytochemical and FTIR Studies on Rhizome of Zingiber officinale Rosc (Gyin) [thesis]. Myanmar: MERAL Portal; 2020

[12] Ahmad B, Rehman MU, Amin I, Arif A, Rasool S, Bhat SA, et al. A review on pharmacological properties of zingerone (4-(4-Hydroxy-3-methoxyphenyl)-2-butanone). The Scientific World Journal. 2015;**2015**:816364

[13] Zadeh JB, Kor NM. Physiological and pharmaceutical effects of ginger (Zingiber officinale Roscoe) as a valuable medicinal plant. European Journal of Experimental Biology. 2014;**4**(1):87-90

[14] Zick SM, Ruffin MT, Djuric Z, Normolle D, Brenner DE. Quantitation of 6-, 8-and 10-gingerols and 6-shogaol in human plasma by high-performance liquid chromatography with electrochemical detection. International Journal of Biomedical Science: IJBS. 2010;**6**(3):233

[15] Semwal RB, Semwal DK, Combrinck S, Viljoen AM. Gingerols

and shogaols: Important nutraceutical principles from ginger. Phytochemistry. 2015;**117**:554-568

[16] Safa O, Hassaniazad M, Farashahinejad M, Davoodian P, Dadvand H, Hassanipour S, et al. Effects of ginger on clinical manifestations and paraclinical features of patients with severe acute respiratory syndrome due to COVID-19: A structured summary of a study protocol for a randomized controlled trial. Trials. 2020;**21**(1):1-2

[17] Shin D, Mukherjee R, Grewe D, Bojkova D, Baek K, Bhattacharya A, et al. Papain-like protease regulates SARS-CoV-2 viral spread and innate immunity. Nature. 2020;**587**(7835):657-662

[18] AlAjmi MF, Azhar A, Owais M, Rashid S, Hasan S, Hussain A, et al. Antiviral potential of some novel structural analogs of standard drugs repurposed for the treatment of COVID-19. Journal of Biomolecular Structure and Dynamics. 2021;**39**(17): 6676-6688

[19] Aldwihi LA, Khan SI, Alamri FF, AlRuthia Y, Alqahtani F, Fantoukh OI, et al. Patients' behavior regarding dietary or herbal supplements before and during COVID-19 in Saudi Arabia. International Journal of Environmental Research and Public Health. 2021;**18**(10):5086

[20] Azam MNK, Al Mahamud R, Hasan A, Jahan R, Rahmatullah M. Some home remedies used for treatment of COVID-19 in Bangladesh. Journal of Medial Plants Studies. 2020;**8**(4):27-32

[21] Wannes WA, Tounsi MS. Can medicinal plants contribute to the cure of Tunisian COVID-19 patients. Journal of Medicinal Plants. 2020;**8**(5):218-226

[22] Orisakwe OE, Orish CN, Nwanaforo EO. Coronavirus disease (COVID-19) and Africa: Acclaimed home remedies. Scientific African. 2020;**10**:e00620

[23] Mesri M, Saber SSE, Godazi M, Shirdel AR, Montazer R, Koohestani HR, et al. The effects of combination of Zingiber officinale and Echinacea on alleviation of clinical symptoms and hospitalization rate of suspected COVID-19 outpatients: A randomized controlled trial. Journal of Complementary and Integrative Medicine. 2021;**18**(4):775-781

[24] Thota SM, Balan V, Sivaramakrishnan V. Natural products as home-based prophylactic and symptom management agents in the setting of COVID-19. Phytotherapy Research. 2020;**34**(12):3148-3167

[25] Tan BK, Vanitha J. Immunomodulatory and antimicrobial effects of some traditional Chinese medicinal herbs: A review. Current Medicinal Chemistry. 2004;**11**(11):1423-1430

[26] Grant KL, Lutz RB. Ginger. American Journal Health-System Pharmacy. 2000;**57**:945-947

[27] Sharma Y. Ginger (Zingiber officinale)-an elixir of life a review. The Pharma Innovation. 2017;**6**(11, Part A):22

[28] Yılmaz N, Seven B, Timur H, Yorgancı A, İnal HA, Kalem MN, et al. Ginger (zingiber officinale) might improve female fertility: A rat model. Journal of the Chinese Medical Association. 2018;**81**(10):905-911

[29] Jittiwat J, Wattanathorn J. Ginger pharmacopuncture improves cognitive impairment and oxidative stress following cerebral ischemia. Journal of Acupuncture and Meridian Studies. 2012;**5**(6):295-300

[30] Mishra RK, Kumar A, Kumar A. Pharmacological activity of Zingiber

officinale. International Journal of Pharmaceutical and Chemical Sciences. 2012;**1**(3):1073-1078

[31] Wynn SG, Luna SPL, Liu H, Xie H, Nan TC, Chien CH. Global acupuncture research: Previously untranslated studies. Studies from Brazil. In: Veterinary Acupuncture. 2nd ed. St. Louis: Mosby; 2001. pp. 53-72

[32] Dehghani I, Mostajeran A, Asghari G. In vitro and in vivo production of gingerols and zingiberene in ginger plant (Zingiber officinale Roscoe). Iranian Journal of Pharmaceutical Sciences. 2011;7(2):117-121

[33] Baliga MS, Haniadka R, Pereira MM, D'Souza JJ, Pallaty PL, Bhat HP, et al. Update on the chemopreventive effects of ginger and its phytochemicals. Critical Reviews in Food Science and Nutrition. 2011;**51**(6):499-523

[34] Ghayur MN, Gilani AH, Afridi MB, Houghton PJ. Cardiovascular effects of ginger aqueous extract and its phenolic constituents are mediated through multiple pathways. Vascular Pharmacology. 2005;**43**(4):234-241

[35] Niksokhan M, Hedarieh N, Maryam N, Masoomeh N. Effect of hydro-alcholic extract of Pimpinella anisum seed on anxiety in male rat. Journal of Gorgan University of Medical Sciences. 2015;**16**(4):28-33

[36] Manju V, Nalini N. Effect of ginger on lipid peroxidation and antioxidant status in 1, 2-dimethyl hydrazine induced experimental colon carcinogenesis. Journal of Biochemical Technology. 2010;**2**(2):161-167

[37] Schwertner HA, Rios DC. High-performance liquid chromatographic analysis of 6-gingerol, 8-gingerol, 10-gingerol, and 6-shogaol in ginger-containing dietary supplements, spices, teas, and beverages. Journal of Chromatography B. 2007;**856**(1-2):41-47

[38] Dugasani S, Pichika MR, Nadarajah VD, Balijepalli MK, Tandra S, Korlakunta JN. Comparative antioxidant and anti-inflammatory effects of [6]-gingerol, [8]-gingerol, [10]-gingerol and [6]-shogaol. Journal of Ethnopharmacology. 2010;**127**(2):515-520

[39] Kundu JK, Surh YJ. Molecular basis of chemoprevention with dietary phytochemicals: Redox-regulated transcription factors as relevant targets. Phytochemistry Reviews. 2009;**8**(2):333-347

[40] Wei CK, Tsai YH, Korinek M, Hung PH, El-Shazly M, Cheng YB, et al. 6-paradol and 6-shogaol, the pungent compounds of ginger, promote glucose utilization in adipocytes and myotubes, and 6-paradol reduces blood glucose in high-fat diet-fed mice. International Journal of Molecular Sciences. 2017;**18**(1):168

[41] Shahrajabian MH, Sun W, Cheng Q. Clinical aspects and health benefits of ginger (Zingiber officinale) in both traditional Chinese medicine and modern industry. Acta agriculturae scandinavica, section b—Soil & Plant Science. 2019;**69**(6):546-556

[42] Kim SO, Kundu JK, Shin YK, Park JH, Cho MH, Kim TY, et al. [6]-Gingerol inhibits COX-2 expression by blocking the activation of p38 MAP kinase and NF-κB in phorbol ester-stimulated mouse skin. Oncogene. 2005;**24**(15):2558-2567

[43] Fadaki F, Modaresi M, Sajjadian I. The effects of ginger extract and diazepam on anxiety reduction in animal model. Indian Journal of

Pharmaceutical Education and Research. 2017;**51**(3):S159-S162

[44] Oludoyin AP, Adegoke SR. Effect of ginger (Zingiber officinale) extracts on blood glucose in normal and streptozotocin-induced diabetic rats. International Journal of Clinical Nutrition. 2014;**2**(2):32-35

[45] El Makawy AI, Ibrahim FM, Mabrouk DM, Ahmed KA, Ramadan MF. Effect of antiepileptic drug (Topiramate) and cold pressed ginger oil on testicular genes expression, sexual hormones and histopathological alterations in mice. Biomedicine & Pharmacotherapy. 2019;**110**:409-419

[46] Sivasothy Y, Chong WK, Hamid A, Eldeen IM, Sulaiman SF, Awang K. Essential oils of Zingiber officinale var. rubrum Theilade and their antibacterial activities. Food Chemistry. 2011;**124**(2):514-517

[47] Park GH, Park JH, Song HM, Eo HJ, Kim MK, Lee JW, et al. Anti-cancer activity of ginger (Zingiber officinale) leaf through the expression of activating transcription factor 3 in human colorectal cancer cells. BMC Complementary and Alternative Medicine. 2014;**14**(1):1-8

[48] Maghbooli M, Golipour F, Moghimi Esfandabadi A, Yousefi M. Comparison between the efficacy of ginger and sumatriptan in the ablative treatment of the common migraine. Phytotherapy Research. 2014;**28**(3):412-415

[49] Lee SH, Cekanova M, Baek SJ. Multiple mechanisms are involved in 6-gingerol-induced cell growth arrest and apoptosis in human colorectal cancer cells. Molecular Carcinogenesis: Published in cooperation with the University of Texas MD Anderson Cancer Center. 2008;**47**(3):197-208

[50] Ryan JL, Heckler CE, Roscoe JA, Dakhil SR, Kirshner J, Flynn PJ, et al. Ginger (Zingiber officinale) reduces acute chemotherapy-induced nausea: A URCC CCOP study of 576 patients. Supportive Care in Cancer. 2012;**20**(7):1479-1489

[51] Pongrojpaw D, Somprasit C, Chanthasenanont A. A randomized comparison of ginger and dimenhydrinate in the treatment of nausea and vomiting in pregnancy. Journal-Medical Association of Thailand. 2007;**90**(9):1703

[52] Khaki A, Farnam A, Badie AD, Nikniaz H. Treatment effects of onion (Allium cepa) and ginger (Zingiber officinale) on sexual behavior of rat after inducing an antiepileptic drug (lamotrigine). Balkan Medical Journal. 2012;**2012**(3):236-242

[53] Hafez DA. Effect of extracts of ginger roots and cinnamon bark on fertility of male diabetic rats. Journal of American Science. 2010;**6**:940-947

[54] Mahassni SH, Bukhari OA. Beneficial effects of an aqueous ginger extract on the immune system cells and antibodies, hematology, and thyroid hormones in male smokers and non-smokers. Journal of Nutrition & Intermediary Metabolism. 2019;**15**:10-17

[55] Atashak S, Peeri M, Azarbayjani MA, Stannard SR. Effects of ginger (Zingiber officinale Roscoe) supplementation and resistance training on some blood oxidative stress markers in obese men. Journal of Exercise Science & Fitness. 2014;**12**(1):26-30

[56] Gagnier JJ, Boon H, Rochon P, Moher D, Barnes J, Bombardier C. Reporting randomised, controlled trials of herbal interventions: An elaborated CONSORT statement. Australian Journal

of Acupuncture and Chinese Medicine. 2006;**1**(1):35-39

[57] Ghayur MN, Gilani AH, Janssen LJ. Ginger attenuates acetylcholine-induced contraction and Ca2+ signalling in murine airway smooth muscle cells. Canadian Journal of Physiology and Pharmacology. 2008;**86**(5):264-271

[58] Aming SN. The Effect of Twice a Day Intake of Ginger Tea on the Blood Pressure of Hypertensive Individuals in Barangay La Victoria, Aurora, Zamboanga Del Sur. Herdin Record#: R09-ZCHRD-12043023205433; Philipine. 2006

[59] Ernst E, Pittler MH. Efficacy of ginger for nausea and vomiting: A systematic review of randomized clinical trials. British Journal of Anaesthesia. 2000;**84**(3):367-371

[60] Knopman DS and Petersen RC. Mild cognitive impairment and mild dementia: A clinical perspective. In: Mayo Clinic Proceedings. Amsterdam, Netherlands: Elsevier. October 2014;**89**(10):1452-1459

[61] Williams KN, Kemper S. Interventions to reduce cognitive decline in aging. Journal of Psychosocial Nursing and Mental Health Services. 2010;**48**(5):42-51

[62] Saenghong N, Wattanathorn J, Muchimapura S, Tongun T, Piyavhatkul N, Banchonglikitkul C, et al. Zingiber officinale improves cognitive function of the middle-aged healthy women. Evidence-based Complementary and Alternative Medicine. 2012;**2012**:383062

[63] Kim HG, Oh MS. Memory-enhancing effect of Mori Fructus via induction of nerve growth factor. British Journal of Nutrition. 2013;**110**(1):86-94

[64] Lim S, Moon M, Oh H, Kim HG, Kim SY, Oh MS. Ginger improves cognitive function via NGF-induced ERK/CREB activation in the hippocampus of the mouse. The Journal of Nutritional Biochemistry. 2014;**25**(10):1058-1065

[65] Moon M, Kim HG, Choi JG, Oh H, Lee PK, Ha SK, et al. 6-Shogaol, an active constituent of ginger, attenuates neuroinflammation and cognitive deficits in animal models of dementia. Biochemical and Biophysical Research Communications. 2014;**449**(1):8-13

[66] Chen RL, Balami JS, Esiri MM, Chen LK, Buchan AM. Ischemic stroke in the elderly: An overview of evidence. Nature Reviews Neurology. 2010;**6**(5):256-265

[67] Arumugam TV, Phillips TM, Cheng A, Morrell CH, Mattson MP, Wan R. Age and energy intake interact to modify cell stress pathways and stroke outcome. Annals of Neurology. 2010;**67**(1):41-52

[68] Sohrabji F, Bake S, Lewis DK. Age-related changes in brain support cells: Implications for stroke severity. Neurochemistry International. 2013;**63**(4):291-301

[69] Wattanathorn J, Jittiwat J, Tongun T, Muchimapura S, Ingkaninan K. Zingiber officinale mitigates brain damage and improves memory impairment in focal cerebral ischemic rat. Evidence-based Complementary and Alternative Medicine. 2010;**2011**:429505

[70] Cloyd J, Hauser W, Towne A, Ramsay R, Mattson R, Gilliam F, et al. Epidemiological and medical aspects of epilepsy in the elderly. Epilepsy Research. 2006;**68**:39-48

[71] Ferlazzo E, Sueri C, Gasparini S, Aguglia U. Challenges in the pharmacological management of epilepsy and its causes in the elderly. Pharmacological Research. 2016;**106**:21-26

[72] Ramsay RE, Rowan AJ, Pryor FM. Special considerations in treating the elderly patient with epilepsy. Neurology. 2004;**62**(5 suppl. 2):S24-S29

[73] Dabbs K, Becker T, Jones J, Rutecki P, Seidenberg M, Hermann B. Brain structure and aging in chronic temporal lobe epilepsy. Epilepsia. 2012;**53**(6):1033-1043

[74] Kondziella D, Bidar A, Urfjell B, Sletvold O, Sonnewald U. The pentylenetetrazole-kindling model of epilepsy in SAMP8 mice: Behavior and metabolism. Neurochemistry International. 2002;**40**(5):413-418

[75] Vishwakarma SL, Pal SC, Kasture VS, Kasture SB. Anxiolytic and antiemetic activity of Zingiber officinale. Phytotherapy Research: An International Journal Devoted to Pharmacological and Toxicological Evaluation of Natural Product Derivatives. 2002;**16**(7):621-626

[76] Hosseini A, Mirazi N. Acute administration of ginger (Zingiber officinale rhizomes) extract on timed intravenous pentylenetetrazol infusion seizure model in mice. Epilepsy Research. 2014;**108**(3):411-419

[77] Ippoushi K, Azuma K, Ito H, Horie H, Higashio H. [6]-Gingerol inhibits nitric oxide synthesis in activated J774. 1 mouse macrophages and prevents peroxynitrite-induced oxidation and nitration reactions. Life Sciences. 2003;**73**(26):3427-3437

Chapter 7

Germicidal and Antineoplastic Activities of Curcumin and Curcumin-Derived Nanoparticles

Lilian Makgoo and Zukile Mbita

Abstract

Curcumin is a major constituent of turmeric and has been shown to have a plethora of health benefits, which include, among many, antimicrobial, anticancer, and reduction of cholesterol. However, it has also been reported that curcumin has less bioaccumulation and is quickly metabolized and cleared from the body. Nanoparticle formulations are known to increase curcumin biocompatibility and targeting. Additionally, the antimicrobial activity of curcumin has been extensively studied and the mechanism of action provides clues for the development of new drugs for drug-resistant microbes. Thus, this chapter will review the biomedical application of curcumin and its nanoformulations against different microbes and other diseases, including cancer.

Keywords: curcumin, nanoparticles, nanomedicine, antimicrobial, antineoplastic

1. Introduction

Curcumin is the major polyphenol component extracted from the rhizomes of *Curcuma longa (C. longa)* [1]. *Curcuma longa* (**Figure 1**) is a perennial herb of the Zingiberaceae family, which is commonly known as turmeric. The rhizome of *C. longa* is rectangular, egg-shaped, pyriform, and has a short branching pattern [1]. Across the globe, this tropical and subtropical plant is widely cultivated in Asia, mostly in India and China [2]. This plant is also cultivated in other regions, including Brazil [3], Nepal [4], Indonesia [5], Jamaica [6], and Pakistan [7, 8]. It was the Polish scientists who first proposed the curcumin structure in 1910 [9]. Curcumin is also known as diferuloylmethane and its IUPAC name is (1E,6E)-1,7-bis(4-hydroxy-3-methoxyphenyl)-1,6-heptadiene-3,5-dione, with a chemical formula of $C_{21}H_{20}O_6$ and, has a molecular weight of 368.38 [10]. Ever since the first isolation of curcumin by two Harvard college scientists, Vogel and Pelletier in 1815 [11], the interest in curcumin and its derivatives have grown steadily and many studies have discovered their biofunctional properties such as anti-inflammatory, antibacterial, anti-tumor and antioxidant activities [12, 13]. Despite being naturally derived, curcumin's derivatives (**Table 1**) are produced by a chemical reaction of aryl-aldehydes with acetylacetone,

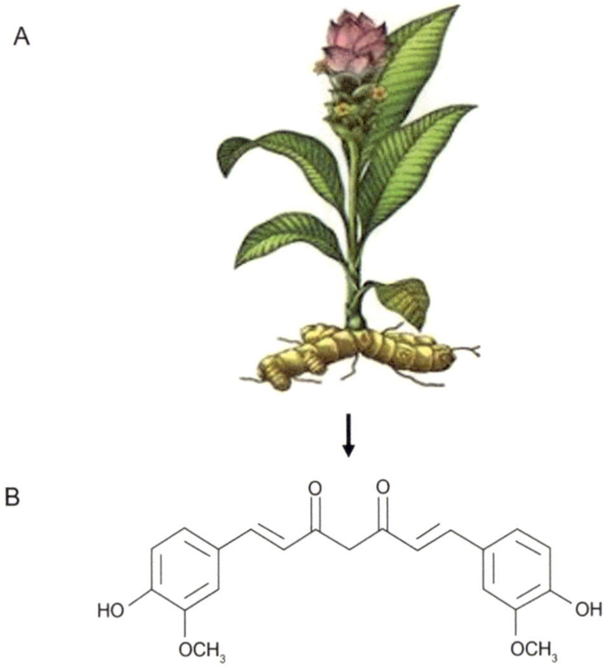

Figure 1.
A & B: Curcuma longa *plant (https://www.istockphoto.com) and the structure of curcumin.*

as a result of this assembly method, multiple chemical analogs can be obtained, for example, compounds in which the middle carbon of the linker (C7) is substituted with an alkyl group [23–25]. A structural modification of curcumin produces compounds with multiple biological activities, such as those useful in the treatment of diabetes, cardiovascular and neurodegenerative diseases [26].

Food and Drug Administration (FDA) has confirmed curcumin to be safe [27]. Several studies have found that curcumin and its derivatives may have anti-inflammatory, antibacterial, antidiabetic, antioxidant, and anticancer benefits (**Table 1**). To possess an anti-inflammatory effect, curcumin blocks the activation of transcription factors, for example, nuclear factor κB (NF-kB), which regulates the expression of pro-inflammatory gene products [28, 29]. The literature on the antibacterial effects of curcumin shows that it damages the cell membranes [30], induces the expression of apoptotic inducers including reactive oxygen species (ROS) [31], and disrupt prokaryotic cell division by inhibiting FtsZ assembly [32]. To relieve diabetic complications, curcumin has been shown to reduce triglycerides levels and inflammation indicators [33]. During inflammation, cyclooxygenase (COX-2) and other pro-inflammatory indicators such NF-κB are produced in greater quantities, these inflammation indicators cause the initiation and development of cancer, thus they are reduced by curcumin [33–35]. Curcumin also prevented the development and progression of cancer by acting as a strong antioxidant agent by regulating the production of ROS, which influence the tumor microenvironment [36]. Additionally, curcumin exerts its anticancer activity by targeting NF-kB, which regulates the expression of proteins such as interleukin (IL)-1, implicated in multiple cell signaling

Curcumin and derivatives/analogues	Structure	Activity	Reference/s
(1E,6E)-1,7-bis(4-hydroxy-3-methoxyphenyl)-1,6-heptadiene-3,5-dione (curcumin)	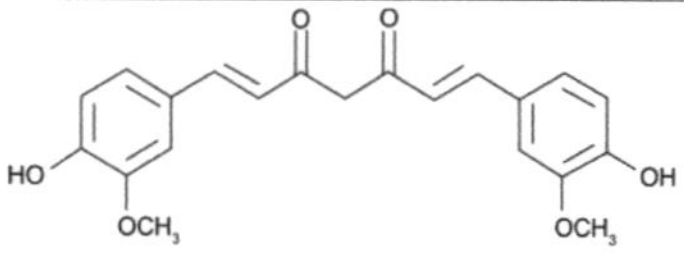	Anti-inflammatory, Antibacterial, Antidiabetic, Anticancer, Antioxidant	[14–17]
(1E, 6Z) 1,7-bis(13-chloro-9-ethylcoumarin-8-yl)-5-hydroxy3-oxohepta-1,4,6-triene (CA2):	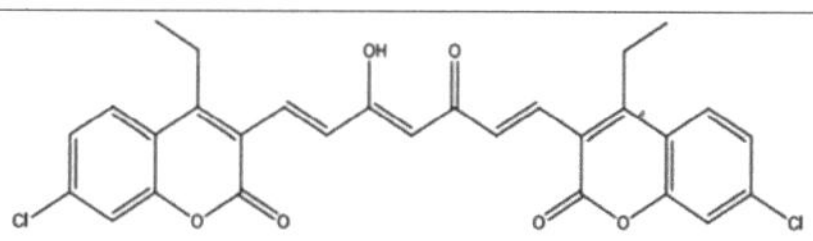	Antibacterial activity	[18]
1,7-Bis(3,4,5-trimethoxy phenyl)-1,6-heptadiene-3,5-dione (L1)	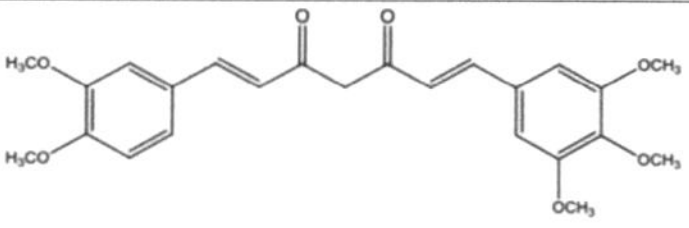	Antifungal activity	[19]
1,7-di(9-anthracenyl)-1,6-heptadiene-3,5-dione (L2)	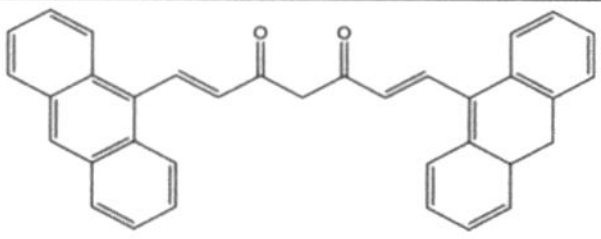	Antifungal activity	[19]
1,5-diphenylpenta-1,4-dien-3-one (3a)	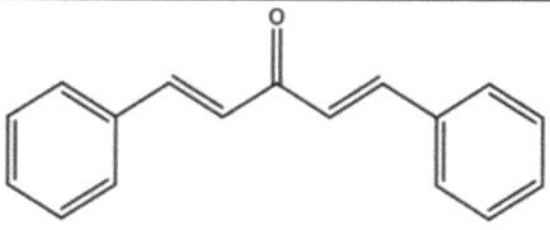	Anti-trichomoniasis activity	[20]
1,5-bis(2-chlorophenyl)penta-1,4-dien-3-one (3e)	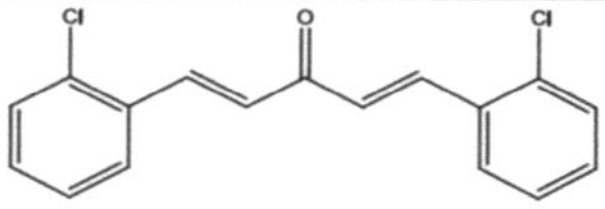	Anti-trichomoniasis activity	[20]
2,6-bis(2-chlorobenzylidene) cyclohexanone (5e)	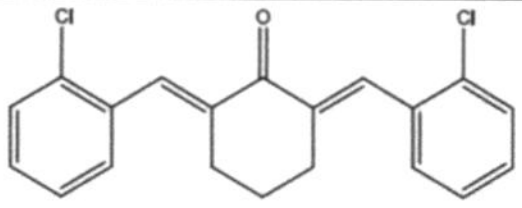	Anti-trichomoniasis activity	[20]
5-bis(4-hydroxy3-methoxybenzylidene)-N-methyl-4-piperidine (PAC)	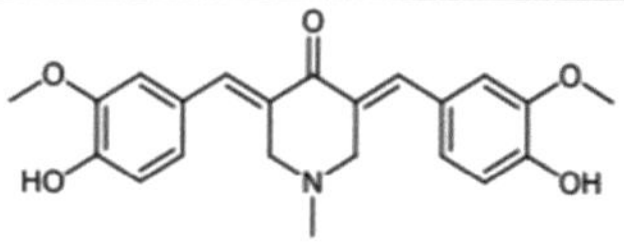	Anti-breast cancer	[21]
1,7-bis-(4-hydroxy-3-methoxyphenyl)-1,6-heptadien-3,5- diene (EAC)	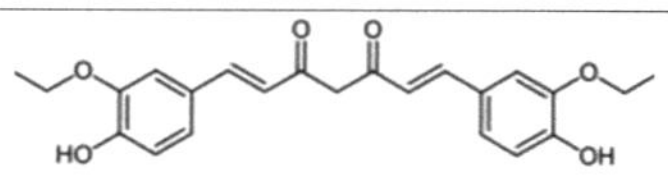	Anti-breast cancer	[21]
Sodium 4-[5-(4-hydroxy-3-methoxyphenyl)-3-oxo-penta-1,4-dienyl]-2-methoxy-phenolate (DM1)	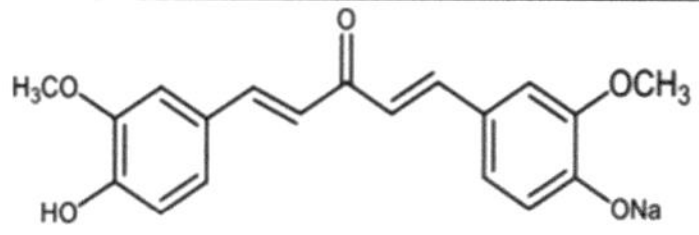	Anti-inflammatory effect	[22]

Table 1.
Structures and activities of curcumin and curcumin derivatives/analogs.

pathways linked to cancer progression and inflammation [25, 37]. Despite its therapeutic potential, curcumin's poor aqueous solubility and low bioavailability remain a challenge [13, 38, 39]. Below we will compare literature on the antibacterial and anticancer activities of curcumin, incorporating nanoformulation as an area that can be explored to fix the therapeutic challenges associated with curcumin.

2. Antibacterial activities of curcumin

The majority of bacteria are not harmful to humans, and some strains even assist in the digestion of food or compete against opportunistic pathogens, but infection by bacteria is one of the most common ailments among humans [40]. Many diseases are connected to bacterial infections, such as inflammatory bowel diseases [41], obesity [42], diabetes [43], liver diseases [44], heart diseases [45], cancers [46], HIV-AIDS [47], and autism [48]. Infections caused by bacteria are largely treated with antibiotics, but the struggle to defeat bacteria continues because bacteria are evolving and manifesting new resistance mechanisms [49]. Curcumin has shown the potential to solve drug resistance issues by inducing antibacterial effect through membrane disruption [30], inducing increased expression of ROS which can promote apoptosis-like response in bacteria [31, 50], and interrupting cell division [32].

Bacteria	Mechanism	Reference/s
Gram-negative bacteria		
Escherichia coli	Inhibit the biofilm formation	[57]
Helicobacter pylori	Growth inhibition Reduce cagA translocation	[58]
Neiserria gonorrhoeae	Reduce cell adherence through the inhibition of NF-kB signaling	[59]
Salmonella sp.	Reduce motility of Salmonella by shortening the length of the flagellar	[60]
Staphylococcus aureus	Inhibit cytokinesis, bacterial proliferation, and cause cell wall damage	[30, 32, 56]
Mycobacterium tuberculosis	Accelerate *Mycobacterium tuberculosis* clearance by promoting antitubercular immunity a	[61]
Gram-positive bacteria		
Bacillus subtilis	Induce membrane permeability Inhibit bacterial cytokinesis	[62, 63]
Bifidobacterium longum BB536	Inhibit cell growth	[64]
Bifidobacterium pseudocatenulatum G4	Inhibit cell growth	[64]
Eenterococcus faecalis	Reduce bacterial growth	[65]
Lactobacillus casei shirota	Reduce bacterial growth	[64]
Lactobacillus acidophilus	Stall bacterial growth	[66]
Sarcina lutea	Phototoxic effect	[67]
Staphylococcus intermedius	Phototoxic effect	[67]

Table 2.
Mechanisms of curcumin on gram-negative and gram-positive bacteria.

Researchers are documenting more evidence about the antibacterial activities of curcumin against a wide range of bacteria [30, 51, 52]. Curcumin has been demonstrated to be potent against both gram-positive and gram-negative bacteria [30, 53]. An example of a gram-positive bacteria, *Staphylococcus aureus* (*S. aureus*), has been demonstrated to be vulnerable to curcumin-mediated inhibition. *Staphylococcus aureus* is a human pathogen that can cause a variety of diseases including infective endocarditis, a feared disease that affect young to middle-aged adults with heart disease [54, 55].

The antibacterial activity of curcumin against *S. aureus* has been thoroughly reviewed by Teow et al. [56]. The *S. aureus* bacteria have developed several mechanisms for evading the human immune system and to resist antibiotic treatment. To salvage *S. aureus* drug resistance, it has been shown that curcumin binds to FtsZ proteins, inhibiting protofilaments assembly, which then inhibits the formation of Z-rings, eliciting the suppression of cytokinesis and bacterial proliferation [32]. Furthermore, the binding of curcumin to peptidoglycans on *S. aureus* cell walls, could cause damage to the cell wall and membrane, hence triggering cell lysis [30, 56]. Mechanisms of curcumin on gram-negative bacteria and gram-positive bacteria are summarized in **Table 2**. In addition to showing its effectiveness as a standalone antibacterial agent, curcumin has also shown marked antibacterial activity when combined with various antibiotics at subinhibitory doses (12.5 and 25 μg/mL) [68, 69]. The collective antibacterial activity of curcumin with antibiotics against methicillin-sensitive *S. aureus* (MSSA) and methicillin-resistant *S. aureus* (MRSA) is well demonstrated by many researchers [68–71]. In tests of *Helicobacter pylori* infection that were done *in-vivo*, mice infected with this bacteria were eradicated by curcumin [72]. In order for curcumin to exert its bactericidal effects, it appears to cause cell membrane damage [30], thus inhibiting bacterial cell division through the improper assembly of the bacterial protofilament, which provides the framework for bacterial cell division apparatus [62, 73].

3. Anticancer activities of curcumin

There were 19.3 million new cancer cases and 10.0 million cancer-related deaths reported in 2020, worldwide [74]. Considering the increasing cancer statistics and the cost of cancer treatments, finding effective and economically viable methods for patients in low- and middle-income countries is crucial. Cancer-related studies showed that curcumin-induced apoptosis and inhibited proliferation in cancer cells through the activation of the mitochondria-mediated pathway [75], ROS generation [76], and the activation of caspase-3 [77]. Other study suggested that curcumin compounds can prevent either the formation or spread of tumor by inducing apoptosis and inhibiting cell proliferation through antiangiogenic effects [78]. Inhibition of tumor invasion by curcumin is mediated by reducing the modification of the matrix metalloproteases (MMPs), the cell surface proteins NF-κβ, TNF-α, cyclooxygenase-2 (COX-2), chemokines, and growth factors (HER-2, EGFR) [79, 80]. In some tumors, curcumin inhibited angiogenesis by suppressing angiogenic cytokines such as IL-6, IL-23, and IL-1β [81].

Cancer and inflammation have a strong relationship, so the anti-inflammatory effects of curcumin would likely result in antitumor effects. According to Pulido-Moran et al. [81], curcumin prevented the development of several types of cancer by reducing the production of mediators of inflammation, such as COX-2 and lipoxygenase 2. The antitumor effect of curcumin has been shown in breast cancer [82], lung cancer [83], leukemia [84], gastric cancer [85], colorectal cancer [86], esophageal cancer [87] and prostate cancer [88]. Curcumin has been shown to regulate key processes involved in cancer development and progression (**Figure 2**).

3.1 Induction of apoptosis

As a form of cell death, apoptosis is a highly regulated physiological process, which removes not only damaged, mutated, aged, and unrepairable cells, but also preserves the integrity and health of the entire organism. Apoptosis imbalances, either excessive or insufficient, may contribute to a variety of diseases including cancer [89, 90]. As a cancer cell growth inhibitor, curcumin modulates multiple cellular signaling pathways such as those that induce apoptosis in several cancers including breast [91], malignant pleural mesothelioma [92], gastric cancer [93], acute lymphoblastic leukemia [94], lung cancer [95], pancreatic cancer [96] and gallbladder carcinoma [97].

Curcumin potentiate apoptosis in cancer due to its ability to induce increased activation of Bax [92], cleavage of poly (ADP-ribose) polymerase (PARP) [92], blocking the PI3K-Akt–mTOR signaling pathway [98, 99], dephosphorylation of Bad [95] and the downregulation of Bcl-2 proteins [96].

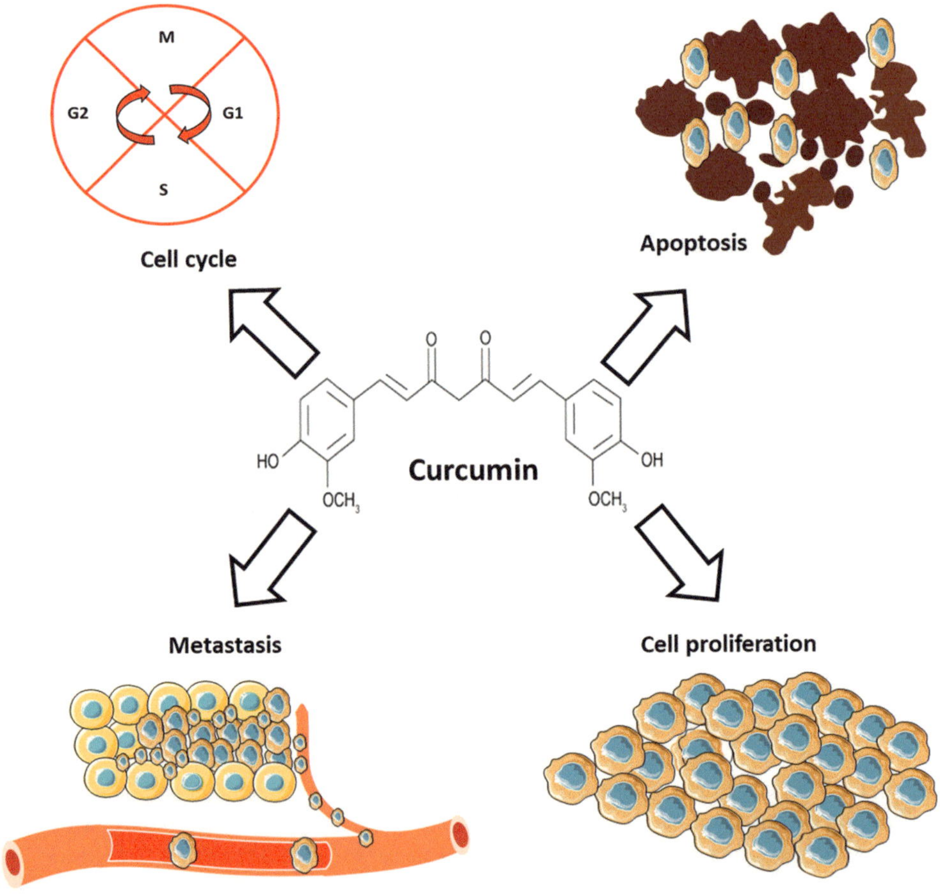

Figure 2.
Cancer processes regulated by curcumin.

3.2 Modulation of cell survival pathways

A number of signaling pathways have been shown to drive unregulated self-renewal and differentiation in cells leading to cancer [100–102]. The antitumor activities of curcumin have been studied extensively; a growing body of evidence indicates that curcumin is involved in the inhibition of growth/proliferation pathways and activation of cell death pathways [103–107]. The fact that curcumin acts through multiple signaling pathways makes it unlikely to develop resistance.

Curcumin regulates multiple cell survival signaling pathways including Wnt/β-catenin pathway [104], NF-κB signaling pathway [105], PI3K/Akt signaling pathway [106], and JAK-STAT3 pathway [107], which regulate different sets of target genes that are involved in cell proliferation, cell survival, and differentiation. The regulation of these cell survival pathways by curcumin has been demonstrated in breast cancer [99, 105], colon cancer [104], bladder cancer [106], lung cancer [108], and liver cancer [109].

3.3 Inhibition of metastasis

Relapse of cancer patients is commonly attributed to cancer invasion and migration, and researchers have been focusing their attention on invasion as an important step in metastasis [110]. Curcumin has shown promising potential for the treatment of cancer by inhibiting metastasis, previous studies have shown that curcumin reduces cancer metastasis by suppressing NF-kB and matrix metalloproteinases (MMPs) expression in cancer animal models [111, 112]. Tumor metastasis is promoted by NF-kB through modulation of cell adhesion molecules including selectins, integrins, and their ligands, NF-kB also induces epithelial-mesenchymal transition, which aids distant metastasis [113]. MMPs also show similar mechanisms by degrading extracellular matrix components resulting in tumor cell migration [114].

Research on the anti-metastasis effect of curcumin continues to pile up, and Sreenivasan et al. [115] showed that curcumin inhibited the metastasis of nasopharyngeal carcinomas (NPCs) by inhibiting miR-125a-5p as a consequence, increasing p53 expression. In prostate cancer, the anti-metastasis effect of curcumin was achieved by decreasing miR-21 and increasing phosphatase and tensin homolog (PTEN) [116]. A recent study of the anti-metastasis effect of curcumin is shown in gastrointestinal cancers, according to this study, curcumin inhibited cell invasion in these cells [117], these results suggest that curcumin inhibits metastasis in cancer by targeting multiple anticancer pathways.

Despite the advantages of curcumin in treating different diseases, the insolubility of curcumin contributes to its poor oral bioavailability and low chemical stability, which limits its application [13, 38, 39]. Moreover, the cellular uptake of curcumin is low, as a result of its hydrophobicity, curcumin penetrates into the cell membrane and binds to the fatty acyl chains of membrane lipids through hydrogen bonds and hydrophobic interactions, resulting in low curcumin levels inside the cytoplasm [56, 118]. These curcumin challenges are resolved by the use of nanoformulations.

4. Curcumin-loaded nanoparticles

Nanoparticles (NPs) have improved the main drawbacks associated with the use of curcumin in biomedical applications, these shortcomings include its rapid metabolism, low solubility, and poor bioavailability, which are considered major obstacles for the treatment of cancer [119], wound healing [120], Alzheimer's disease [121], epilepsy [122] ischemia [123], and inflammatory diseases [124]. The delivery of therapeutic concentrations of curcumin using nanoparticles is emerging as one of the most useful alternatives to treat different diseases including cancer and microbial infections (**Figure 3**). **Table 3** summarizes the therapeutic potential of curcumin-loaded nanoparticles on different diseases.

4.1 Curcumin-loaded nanoparticles and their antibacterial activities

According to a study by Tyagi et al. [30], curcumin has greater effectiveness in controlling both gram-positive and gram-negative bacteria. Despite curcumin having potential antibacterial properties, its low solubility, low stability, and low bioavailability remain a debate [142]. Curcumin nanoparticles exhibit better biological activity, solubility, and stability than all other forms of curcumin [143]. When evaluated, it was found that the curcumin particles are smaller, which enhanced their toxicity and sensitization in bacterial cells, compared to curcumin alone [144].

Furthermore, an evaluation of curcumin nanoparticle's effectiveness in inhibiting bacteria like *Shigella dysenteriae*, *Staphylococcus aureus*, *Escherichia coli*, and *Streptococcus pneumonia* was observed to be greater than amoxicillin, a commercial

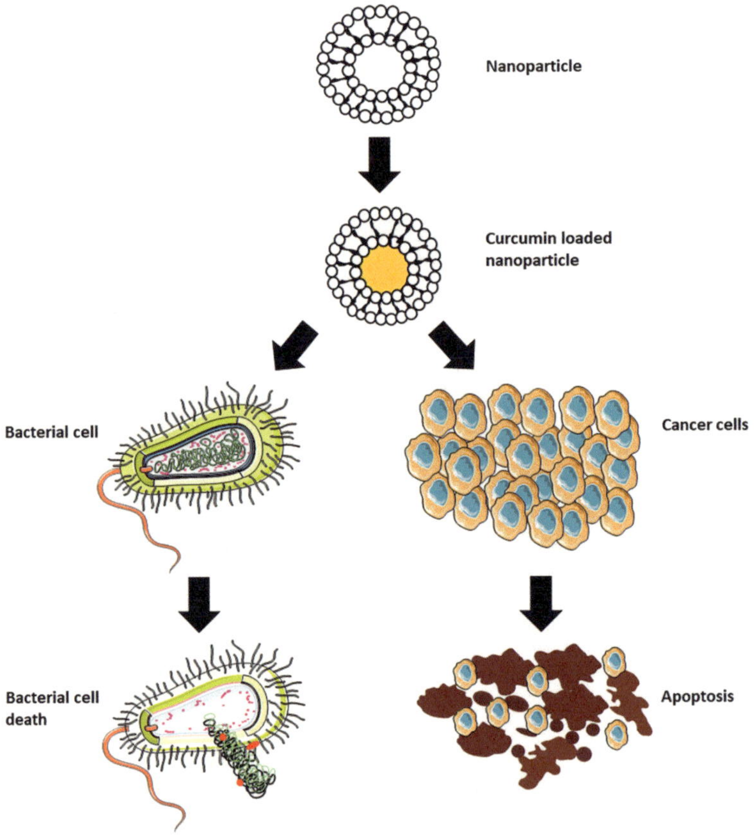

Figure 3.
Curcumin-loaded nanoparticles induce cell death in both cancer and bacterial cells.

Nanoparticle	Disease	Outcome	Reference/s
Gold	Prostate cancer Colorectal cancer Renal cancer	Improved solubility Augmented antioxidant and anticancer effects	[125–127]
Magnetic	Inflammatory cells Breast cancer	Higher drug encapsulation Higher stability, and loading efficiency Anticancer effects Active protection against inflammatory agents	[128–131]
Silver	Bacterial infections Wound healing	Antibacterial activity	[132, 133]
Chitosan	Malaria Diabetic wound healing	Antimalarial activity Better bioavailability	[134, 135]
Solid lipid	Inflammation Breast cancer Cerebral ischemia	Increased solubility Anti-inflammatory Antitumor	[136–139]
Nanogel	Breast cancer Skin cancer	Induced cytotoxicity and apoptosis	[140, 141]

Table 3.
Summary of the activities of curcumin-loaded nanoparticles for the treatment of different diseases.

antibiotic [52]. In order to manifest antibacterial properties, curcumin-loaded nanoparticles attach to the cell wall of the bacterial cell, break it, and penetrate inside the cell, disrupting the structure of cellular organelles [143]. Since curcumin-loaded nanoparticles are effective on the broad spectrum of microorganisms and human cancer cell lines, therefore targeting curcumin-loaded nanoparticles for therapeutic purpose is a promising strategy.

4.2 Curcumin-loaded nanoparticles and their anticancer activities

Over the years, different nanoformulations have been investigated in order to enhance the delivery of curcumin to tumor sites [128, 145, 146]. Different nanoparticle-based approaches have been explored, such as solid-lipid microparticles based on bovine serum albumin [147, 148], encapsulation in liposomes [149], and chitosan [150]. These nanoparticles are tailored in a precise dimension for the purpose of increasing absorption and permeation, which then result in more bio-distribution and longer circulation in the body [151]. Nanoformulations are used primarily for enhancing the solubility of curcumin in water [152]. To enhance solubility, curcumin is prepared using pH-driven loading method, in this method, hydrophobic phytochemicals such as curcumin are deprotonated and dissolved under alkaline conditions to overcome solubility challenges [152, 153].

Preferably, curcumin nanoformulation would exhibit increased anticancer activity over free curcumin, while remaining nontoxic to normal cells. Chabib et al. [154] compared the anticancer activity of pure curcumin with curcumin-loaded nanoparticles, and found that curcumin-loaded nanoparticles were more effective than curcumin on its own against breast cancer cells T47D. A study by Bisht et al. [155] had previously demonstrated that pancreatic cancer can be effectively treated with polymer-based curcumin-loaded nanoparticles, which induced apoptosis and obstructed the activation of NFκB in BxPC3 pancreatic cancer cells. Recently, the use of curcumin-loaded nanoparticles in combination with anticancer drugs have been shown to enhance their chemotherapeutic effect in ovarian carcinoma by inhibiting proliferation via modulation of JAK/STAT3 and PI3K/Akt signaling pathways [156].

Curcumin-loaded nanoparticles showed increased anticancer effect in lung cancer [157], prostate cancer [158], breast cancer [154], colon cancer [159], brain cancer [160] and oral cancer [161]. Additionally, curcumin-loaded nanoparticles have been shown to interact with plasma proteins, providing a new platform for improving cancer treatment [162]. Based on these results, it is not surprising that the usage of curcumin-loaded nanoparticles is gaining momentum in anticancer therapeutics [163].

4.3 Strategies to improve curcumin nanoparticles

Numerous *in vitro* and *in vivo* studies have shown that nanoparticles may enhance the anticancer effects of curcumin [164, 165]. However, there are still some concerns about the cost, safety, side-effects, and long-term toxicity of curcumin-loaded nanoparticles, leading to the development of a new field of study called nanotoxicology [166].

To improve curcumin nanoparticles, curcumin-loaded nanoparticles should be tested in a larger population to determine their toxicity and efficacy. Furthermore, clinical trials are necessary to evaluate their anticancer activities and determine side effects and toxicity in human subjects [167]. The safety concerns associated with nanomedicine-based delivery systems include neuroinflammation, excitotoxicity, and DNA damage [168]. Although they are methods that are currently explored to reduce the toxicity of nanoparticles [169], developing DNA/RNA nano-carriers to eliminate cancer cells can be a promising plan of action.

5. Conclusion

The safety profile of curcumin is exceptional, and it has multiple health benefits including anti-inflammatory, antioxidant, antitumor, antibacterial, and anti-diabetic properties. However, the poor stability in the body fluids, rapid clearance, and low aqueous solubility limit curcumin's clinical use. Nano-based drug delivery systems are currently opening a new world of possibilities for solving these problems. Nanoparticles have been shown to improve the solubility and stability of some substances including curcumin and amend its curative index. Therefore, targeting curcumin-loaded nanoparticles for therapeutic purpose is a promising strategy.

Acknowledgements

The authors wish to thank the University of Limpopo Office for Research Development and Administration.

Conflict of interest

The authors declare no conflict of interest.

Notes/thanks/other declarations

The authors wish to thank the Department of Biochemistry, Microbiology and Biotechnology (University of Limpopo) for support.

Author details

Lilian Makgoo and Zukile Mbita*
Department of Biochemistry, Microbiology and Biotechnology, University of Limpopo, Polokwane, South Africa

*Address all correspondence to: zukile.mbita@ul.ac.za

References

[1] Alibeiki F, Jafari N, Karimi M, Dogaheh HP. Potent anti-cancer effects of less polar curcumin analogues on gastric adenocarcinoma and esophageal squamous cell carcinoma cells. Scientific Reports. 2017;**7**(1):1-9

[2] Nawaz A. Curcumin: A natural product of biological importance. Gomal University Journal of Research. 2011;**27**(1):7-14

[3] Pino JA, Fon-Fay FM, Pérez JC, Falco AS, Hernández I, Rodeiro I, et al. Chemical composition and biological activities of essential oil from turmeric (Curcuma longa L.) rhizomes grown in Amazonian Ecuador. Revista CENIC. Ciencias Químicas. 2018;**49**(1):1-8

[4] Devkota L, Rajbhandari M. Composition of essential oils in turmeric rhizome. Nepal Journal of Science and Technology. 2015;**16**(1):87-94

[5] Hong SL, Lee GS, Syed Abdul Rahman SN, Ahmed Hamdi OA, Awang K, Aznam Nugroho N, et al. Essential oil content of the rhizome of Curcuma purpurascens Bl.(Temu Tis) and its antiproliferative effect on selected human carcinoma cell lines. The Scientific World Journal. 2014;**2014**:397430

[6] Green CE, Mitchell SA. The effects of blanching, harvest time and location (with a minor look at postharvest blighting) on oleoresin yields, percent curcuminoids and levels of antioxidant activity of turmeric (Curcuma longa) rhizomes grown in Jamaica. Modern Chemistry and Applications. 2014;**2**(140):2-9

[7] Naz S, Ilyas S, Parveen Z, Javed S. Chemical analysis of essential oils from turmeric (Curcuma longa) rhizome through GC-MS. Asian Journal of Chemistry. 2010;**22**(4):3153

[8] Naz S, Ilyas S, Jabeen S, Parveen Z. Composition and antibacterial activity of the essential oil from the rhizome of turmeric (Curcuma longa L.). Asian J. Chem. 2011;**23**(4):1639-1642

[9] Miłobędzka J. v. Kostanecki S, Lampe V. Zur kenntnis des curcumins. Berichte der deutschen chemischen Gesellschaft. 1910;**43**(2):2163-2170

[10] Giordano A, Tommonaro G. Curcumin and cancer. Nutrients. 2019;**11**(10):2376

[11] Vogel A, Pelletier J. Examen chimique de la racine de Curcuma. Journal of Pharmaceutics. 1815;**1**:289-300

[12] Aggarwal BB, Deb L, Prasad S. Curcumin differs from tetrahydrocurcumin for molecular targets, signaling pathways and cellular responses. Molecules. 2015;**20**(1):185-205

[13] Nagahama K, Utsumi T, Kumano T, Maekawa S, Oyama N, Kawakami J. Discovery of a new function of curcumin which enhances its anticancer therapeutic potency. Scientific Reports. 2016;**6**(1):1-4

[14] Aggarwal BB, Sung B. Pharmacological basis for the role of curcumin in chronic diseases: An age-old spice with modern targets. Trends in Pharmacological Sciences. 2009; **30**(2):85-94

[15] Dai Q, Di Zhou LX, Song X. Curcumin alleviates rheumatoid arthritis-induced inflammation and synovial hyperplasia by targeting mTOR pathway in rats. Drug Design, Development and Therapy. 2018;**12**:4095

[16] Adamczak A, Ożarowski M, Karpiński TM. Curcumin, a natural antimicrobial agent with strain-specific activity. Pharmaceuticals. 2020;**13**(7):153

[17] Den Hartogh DJ, Gabriel A, Tsiani E. Antidiabetic properties of curcumin II: Evidence from in vivo studies. Nutrients. 2020;**12**(1):58

[18] Oglah MK, Mustafa YF. Curcumin analogs: Synthesis and biological activities. Medicinal Chemistry Research. 2020;**29**(3):479-486

[19] Thomachan S, Sindhu S, John VD. Synthesis, characterization, antibacterial, antifungal and cytotoxic activity of curcuminoid analogues with trisubstituted phenyl and anthracenyl ring and their zinc (II), copper (II) and vanadyl (IV) chelates. International Journal of Pharmaceutical Chemistry. 2016;**6**:78-86

[20] da Silva CC, Pacheco BS. das Neves RN, Alves MS, Sena-lopes Â, Moura S, Borsuk S, de Pereira CM. Antiparasitic activity of synthetic curcumin monocarbonyl analogues against trichomonas vaginalis. Biomedicine & Pharmacotherapy. 2019;**111**:367-377

[21] Al-Hujaily EM, Mohamed AG, Al-Sharif I, Youssef KM, Manogaran PS, Al-Otaibi B, et al. PAC, a novel curcumin analogue, has anti-breast cancer properties with higher efficiency on ER-negative cells. Breast Cancer Research and Treatment. 2011;**128**(1):97-107

[22] Paulino N, Paulino AS, Diniz SN, de Mendonça S, Gonçalves ID, Flores FF, et al. Evaluation of the anti-inflammatory action of curcumin analog (DM1): Effect on iNOS and COX-2 gene expression and autophagy pathways. Bioorganic & Medicinal Chemistry. 2016;**24**(8):1927-1935

[23] Chen WF, Deng SL, Zhou B, Yang L, Liu ZL. Curcumin and its analogues as potent inhibitors of low density lipoprotein oxidation: H-atom abstraction from the phenolic groups and possible involvement of the 4-hydroxy-3-methoxyphenyl groups. Free Radical Biology and Medicine. 2006;**40**(3):526-535

[24] Borik RM, Fawzy NM, Abu-Bakr SM, Aly MS. Design, synthesis, anticancer evaluation and docking studies of novel heterocyclic derivatives obtained via reactions involving curcumin. Molecules. 2018;**23**(6):1398

[25] Tomeh MA, Hadianamrei R, Zhao X. A review of curcumin and its derivatives as anticancer agents. International Journal of Molecular Sciences. 2019;**20**(5):1033

[26] Mbese Z, Khwaza V, Aderibigbe BA. Curcumin and its derivatives as potential therapeutic agents in prostate, colon and breast cancers. Molecules. 2019;**24**(23):4386

[27] Sharma RA, Gescher AJ, Steward WP. Curcumin: The story so far. European Journal of Cancer. 2005;**41**(13):1955-1968

[28] Li Q, Sun J, Mohammadtursun N, Wu J, Dong J, Li L. Curcumin inhibits cigarette smoke-induced inflammation via modulating the PPARγ-NF-κB signaling pathway. Food & Function. 2019;**10**(12):7983-7994

[29] Zhu T, Chen Z, Chen G, Wang D, Tang S, Deng H, et al. Curcumin attenuates asthmatic airway inflammation and mucus hypersecretion involving a PPARγ-dependent NF-κB signaling pathway in vivo and in vitro. Mediators of Inflammation. 2019;**3**:2019

[30] Tyagi P, Singh M, Kumari H, Kumari A, Mukhopadhyay K.

Bactericidal activity of curcumin I is associated with damaging of bacterial membrane. PLoS One. 2015;**10**(3):e0121313

[31] Yun DG, Lee DG. Antibacterial activity of curcumin via apoptosis-like response in Escherichia coli. Applied Microbiology and Biotechnology. 2016;**100**(12):5505-5514

[32] Kaur S, Modi NH, Panda D, Roy N. Probing the binding site of curcumin in Escherichia coli and Bacillus subtilis FtsZ–a structural insight to unveil antibacterial activity of curcumin. European Journal of Medicinal Chemistry. 2010;**45**(9):4209-4214

[33] Adibian M, Hodaei H, Nikpayam O, Sohrab G, Hekmatdoost A, Hedayati M. The effects of curcumin supplementation on high-sensitivity C-reactive protein, serum adiponectin, and lipid profile in patients with type 2 diabetes: A randomized, double-blind, placebo-controlled trial. Phytotherapy Research. 2019;**33**(5):1374-1383

[34] Mantovani A. Molecular pathways linking inflammation and cancer. Current Molecular Medicine. 2010;**10**(4):369-373

[35] Mohamed SI, Jantan I, Haque MA. Naturally occurring immunomodulators with antitumor activity: An insight on their mechanisms of action. International Immunopharmacology. 2017;**50**:291-304

[36] Nakamae I, Morimoto T, Shima H, Shionyu M, Fujiki H, Yoneda-Kato N, et al. Curcumin derivatives verify the essentiality of ROS upregulation in tumor suppression. Molecules. 2019;**24**(22):4067

[37] Sethi G, Tergaonkar V. Potential pharmacological control of the NF-κB pathway. Trends in Pharmacological Sciences. 2009;**30**(6):313-321

[38] Shen L, Liu CC, An CY, Ji HF. How does curcumin work with poor bioavailability? Clues from experimental and theoretical studies. Scientific Reports. 2016;**6**(1):1

[39] Fadus MC, Lau C, Bikhchandani J, Lynch HT. Curcumin: An age-old anti-inflammatory and anti-neoplastic agent. Journal of Traditional and Complementary Medicine. 2017;7(3):339-346

[40] Lynch SV, Pedersen O. The human intestinal microbiome in health and disease. New England Journal of Medicine. 2016;**375**(24):2369-2379

[41] Jostins L, Ripke S, Weersma RK, Duerr RH, McGovern DP, Hui KY, et al. Host–microbe interactions have shaped the genetic architecture of inflammatory bowel disease. Nature. 2012;**491**(7422):119-124

[42] Leung J, Burke B, Ford D, Garvin G, Korn C, Sulis C, et al. Possible association between obesity and Clostridium difficile infection. Emerging Infectious Diseases. Nov 2013;**19**(11):1791

[43] Hara N, Alkanani AK, Ir D, Robertson CE, Wagner BD, Frank DN, et al. The role of the intestinal microbiota in type 1 diabetes. Clinical Immunology. 2013;**146**(2):112-119

[44] Sargenti K, Prytz H, Nilsson E, Kalaitzakis E. Predictors of mortality among patients with compensated and decompensated liver cirrhosis: The role of bacterial infections and infection-related acute-on-chronic liver failure. Scandinavian Journal of Gastroenterology. 2015;**50**(7):875-883

[45] Izadi M, Fazel M, Sharubandi SH, Saadat SH, Farahani MM, Nasseri MH, et al. Helicobacter species in the atherosclerotic plaques of patients with

coronary artery disease. Cardiovascular Pathology. 2012;**21**(4):307-311

[46] Weir TL, Manter DK, Sheflin AM, Barnett BA, Heuberger AL, Ryan EP. Stool microbiome and metabolome differences between colorectal cancer patients and healthy adults. PLoS One. 2013;**8**(8):e70803

[47] Gori A, Rizzardini G, Van't Land B, Amor KB, Van Schaik J, Torti C, et al. Specific prebiotics modulate gut microbiota and immune activation in HAART-naive HIV-infected adults: Results of the "COPA" pilot randomized trial. Mucosal Immunology. 2011;**4**(5):554-563

[48] De Angelis M, Francavilla R, Piccolo M, De Giacomo A, Gobbetti M. Autism spectrum disorders and intestinal microbiota. Gut Microbes. 2015;**6**(3):207-213

[49] Fair RJ, Tor Y. Antibiotics and bacterial resistance in the 21st century. Perspectives in Medicinal Chemistry. 2014;**6**:PMC-S14459

[50] Dwyer DJ, Camacho DM, Kohanski MA, Callura JM, Collins JJ. Antibiotic-induced bacterial cell death exhibits physiological and biochemical hallmarks of apoptosis. Molecular Cell. 2012;**46**(5):561-572

[51] Gupta SC, Patchva S, Koh W, Aggarwal BB. Discovery of curcumin, a component of golden spice, and its miraculous biological activities. Clinical and Experimental Pharmacology and Physiology. 2012;**39**(3):283-299

[52] Sharifi S, Fathi N, Memar MY, Hosseiniyan Khatibi SM, Khalilov R, Negahdari R, et al. Anti-microbial activity of curcumin nanoformulations: New trends and future perspectives. Phytotherapy Research. 2020;**34**(8): 1926-1946

[53] Zorofchian Moghadamtousi S, Abdul Kadir H, Hassandarvish P, Tajik H, Abubakar S, Zandi K. A review on antibacterial, antiviral, and antifungal activity of curcumin. BioMed Research International. 2014;**2014**:186864

[54] Seckeler MD, Hoke TR. The worldwide epidemiology of acute rheumatic fever and rheumatic heart disease. Clinical Epidemiology. 2011;**3**:67

[55] Tong SY, Davis JS, Eichenberger E, Holland TL, Fowler VG Jr. Staphylococcus aureus infections: Epidemiology, pathophysiology, clinical manifestations, and management. Clinical Microbiology Reviews. 2015;**28**(3):603-661

[56] Teow SY, Liew K, Ali SA, Khoo AS, Peh SC. Antibacterial action of curcumin against Staphylococcus aureus: A brief review. Journal of Tropical Medicine. 2016;**2016**:2853045

[57] Packiavathy IA, Priya S, Pandian SK, Ravi AV. Inhibition of biofilm development of uropathogens by curcumin–an anti-quorum sensing agent from Curcuma longa. Food Chemistry. 2014;**148**:453-460

[58] Ray AK, Luis PB, Mishra SK, Barry DP, Asim M, Pandey A, et al. Curcumin oxidation is required for inhibition of helicobacter pylori growth, translocation and phosphorylation of cag a. Frontiers in Cellular and Infection Microbiology. 2021;**11**:765842

[59] Wessler S, Muenzner P, Meyer TF, Naumann M. The anti-inflammatory compound curcumin inhibits Neisseria gonorrhoeae-induced NF-κB signaling, release of pro-inflammatory cytokines/ chemokines and attenuates adhesion

in late infection. Journal of Biological Chemistry. 2005;**386**:481-490

[60] Marathe SA, Balakrishnan A, Negi VD, Sakorey D, Chandra N, Chakravortty D. Curcumin reduces the motility of salmonella enterica serovar typhimurium by binding to the flagella, thereby leading to flagellar fragility and shedding. Journal of Bacteriology. 2016;**198**(13):1798-1811

[61] Tousif S, Singh DK, Mukherjee S, Ahmad S, Arya R, Nanda R, et al. Nanoparticle-formulated curcumin prevents posttherapeutic disease reactivation and reinfection with mycobacterium tuberculosis following isoniazid therapy. Frontiers in Immunology. 2017;**8**:739

[62] Rai D, Singh JK, Roy N, Panda D. Curcumin inhibits FtsZ assembly: An attractive mechanism for its antibacterial activity. Biochemical Journal. 2008;**410**(1):147-155

[63] Morão LG, Polaquini CR, Kopacz M, Torrezan GS, Ayusso GM, Dilarri G, et al. A simplified curcumin targets the membrane of Bacillus subtilis. Microbiology Open. 2019;**8**(4):e00683

[64] Jazayeri SD, Mustafa S, Manap MY, Ali AM, Ismail A, Faujan NH, et al. Survival of bifidobacteria and other selected intestinal bacteria in TPY medium supplemented with curcumin as assessed in vitro. International Journal of Probiotics and Prebiotics. 2009;**4**:15-22

[65] Sainudeen S, Nair VS, Zarbah M, Abdulla AM, Najeeb CM, Ganapathy S. Can herbal extracts serve as antibacterial root canal irrigating solutions? Antimicrobial efficacy of Tylophora indica, curcumin longa, Phyllanthus amarus, and sodium hypochlorite on enterococcus faecalis biofilms formed on tooth substrate: In vitro study. Journal of Pharmacy & Bioallied Sciences. 2020;**12**(Suppl. 1):S423-S429

[66] Araújo NC, De Menezes RF, Carneiro VS, dos Santos-Neto AP, Fontana CR, Bagnato VS, et al. Photodynamic inactivation of cariogenic pathogens using curcumin as photosensitizer. Photomedicine and Laser Surgery. 2017;**35**(5):259-263

[67] Haukvik T, Bruzell E, Kristensen S, Tønnesen HH. Photokilling of bacteria by curcumin in different aqueous preparations. Studies on curcumin and curcuminoids XXXVII. Die Pharmazie-An International Journal of Pharmaceutical Sciences. 200;**64**(10):666-673

[68] Moghaddam KM, Iranshahi M, Yazdi MC, Shahverdi AR. The combination effect of curcumin with different antibiotics against Staphylococcus aureus. International Journal of Green Pharmacy. 2009;**1**:3

[69] Teow SY, Ali SA. Synergistic antibacterial activity of curcumin with antibiotics against Staphylococcus aureus. Pakistan Journal of Pharmaceutical Sciences. 2015;**28**(6):2109-2114

[70] Mun SH, Joung DK, Kim YS, Kang OH, Kim SB, Seo YS, et al. Synergistic antibacterial effect of curcumin against methicillin-resistant Staphylococcus aureus. Phytomedicine. 2013;**20**(8-9):714-718

[71] Sasidharan NK, Sreekala SR, Jacob J, Nambisan B. In vitro synergistic effect of curcumin in combination with third generation cephalosporins against bacteria associated with infectious diarrhea. BioMed Research International. 2014;**1**:2014

[72] De R, Kundu P, Swarnakar S, Ramamurthy T, Chowdhury A, Nair GB,

et al. Antimicrobial activity of curcumin against Helicobacter pylori isolates from India and during infections in mice. Antimicrobial Agents and Chemotherapy. 2009;**53**(4):1592-1597

[73] Guan F, Yu J, Yu J, Liu Y, Li Y, Feng XH, et al. Lateral interactions between protofilaments of the bacterial tubulin homolog FtsZ are essential for cell division. eLife. 2018;**7**:e35578

[74] Ferlay J, Colombet M, Soerjomataram I, Parkin DM, Piñeros M, Znaor A, et al. Cancer statistics for the year 2020: An overview. International Journal of Cancer. 2021;**5**:778-789

[75] Guo LD, Chen XJ, Hu YH, Yu ZJ, Wang D, Liu JZ. Curcumin inhibits proliferation and induces apoptosis of human colorectal cancer cells by activating the mitochondria apoptotic pathway. Phytotherapy Research. 2013;**27**(3):422-430

[76] Agarwal A, Kasinathan A, Ganesan R, Balasubramanian A, Bhaskaran J, Suresh S, et al. Curcumin induces apoptosis and cell cycle arrest via the activation of reactive oxygen species–independent mitochondrial apoptotic pathway in Smad4 and p53 mutated colon adenocarcinoma HT29 cells. Nutrition Research. 2018;**51**:67-81

[77] Lou S, Wang Y, Yu Z, Guan K, Kan Q. Curcumin induces apoptosis and inhibits proliferation in infantile hemangioma endothelial cells via downregulation of MCL-1 and HIF-1α. Medicine. Feb 2018;**97**(7):e10654

[78] Prasad S, Gupta SC, Tyagi AK, Aggarwal BB. Curcumin, a component of golden spice: From bedside to bench and back. Biotechnology Advances. 2014;**32**(6):1053-1064

[79] Niedzwiecki A, Roomi MW, Kalinovsky T, Rath M. Anticancer efficacy of polyphenols and their combinations. Nutrients. 2016;**8**(9):552

[80] Qadir MI, Naqvi ST, Muhammad SA, Qadir M, Naqvi ST. Curcumin: A polyphenol with molecular targets for cancer control. Asian Pacific Journal of Cancer Prevention. 2016;**17**(6):2735-2739

[81] Pulido-Moran M, Moreno-Fernandez J, Ramirez-Tortosa C, Ramirez-Tortosa M. Curcumin and health. Molecules. 2016;**21**(3):264

[82] Kim JM, Noh EM, Kwon KB, Kim JS, You YO, Hwang JK, et al. Curcumin suppresses the TPA-induced invasion through inhibition of PKCα-dependent MMP-expression in MCF-7 human breast cancer cells. Phytomedicine. 2012;**19**(12):1085-1092

[83] Jin H, Qiao F, Wang Y, Xu Y, Shang Y. Curcumin inhibits cell proliferation and induces apoptosis of human non-small cell lung cancer cells through the upregulation of miR-192-5p and suppression of PI3K/Akt signaling pathway. Oncology Reports. 2015;**34**(5):2782-2789

[84] Yang CW, Chang CL, Lee HC, Chi CW, Pan JP, Yang WC. Curcumin induces the apoptosis of human monocytic leukemia THP-1 cells via the activation of JNK/ERK pathways. BMC Complementary and Alternative Medicine. 2012;**12**(1):1-8

[85] Thao DT, Phuong DT, Hanh TT, Thao NP, Cuong NX, Nam NH, et al. Two new neoclerodane diterpenoids from Scutellaria barbata D. Don growing in Vietnam. Journal of Asian Natural Products Research. 2014;**16**(4):364-369

[86] Lim TG, Lee SY, Huang Z, Chen H, Jung SK, Bode AM, et al. Curcumin suppresses proliferation of colon

cancer cells by targeting CDK2. Cancer Prevention Research. 2014;**7**(4):466-474

[87] Li XJ, Li YZ, Jin CT, Fan J, Li HJ. Curcumin induces apoptosis by PTEN/PI3K/AKT pathway in EC109 cells. Zhongguo ying yong sheng li xue za zhi= Zhongguo yingyong shenglixue zazhi. Chinese Journal of Applied Physiology. 2015;**31**(2):174-177

[88] Du Y, Long Q, Zhang L, Shi Y, Liu X, Li X, et al. Curcumin inhibits cancer-associated fibroblast-driven prostate cancer invasion through MAOA/mTOR/HIF-1α signaling. International Journal of Oncology. 2015;**47**(6):2064-2072

[89] Favaloro B, Allocati N, Graziano V, Di Ilio C, De Laurenzi V. Role of apoptosis in disease. Aging (Albany NY). 2012;**4**(5):330

[90] Thapa S, Rather RA, Singh SK, Bhagat M. Insights into the role of defective apoptosis in cancer pathogenesis and therapy. Available from: https://www.intechopen.com/online-first/76842

[91] Lv ZD, Liu XP, Zhao WJ, Dong Q, Li FN, Wang HB, et al. Curcumin induces apoptosis in breast cancer cells and inhibits tumor growth in vitro and in vivo. International Journal of Clinical and Experimental Pathology. 2014;**7**(6):2818

[92] Zhang C, Hao Y, Wu L, Dong X, Jiang N, Cong B, et al. Curcumin induces apoptosis and inhibits angiogenesis in murine malignant mesothelioma. International Journal of Oncology. 2018;**53**(6):2531-2541

[93] Li W, Zhou Y, Yang J, Li H, Zhang H, Zheng P. Curcumin induces apoptotic cell death and protective autophagy in human gastric cancer cells. Oncology Reports. 2017;**37**(6):3459-3466

[94] Kuttikrishnan S, Siveen KS, Prabhu KS, Khan AQ, Ahmed EI, Akhtar S, et al. Curcumin induces apoptotic cell death via inhibition of PI3-kinase/AKT pathway in B-precursor acute lymphoblastic leukemia. Frontiers in Oncology. 2019;**9**:484

[95] Endo H, Inoue I, Masunaka K, Tanaka M, Yano M. Curcumin induces apoptosis in lung cancer cells by 14-3-3 protein-mediated activation of bad. Bioscience, Biotechnology, and Biochemistry. 2020;**84**(12):2440-2447

[96] Zhu Y, Bu S. Curcumin induces autophagy, apoptosis, and cell cycle arrest in human pancreatic cancer cells. Evidence-Based Complementary and Alternative Medicine. 2017;**1**: 2017

[97] Liu TY, Tan ZJ, Jiang L, Gu JF, Wu XS, Cao Y, et al. Curcumin induces apoptosis in gallbladder carcinoma cell line GBC-SD cells. Cancer Cell International. 2013;**13**(1):1-9

[98] Fu H, Wang C, Yang D, Wei Z, Xu J, Hu Z, et al. Curcumin regulates proliferation, autophagy, and apoptosis in gastric cancer cells by affecting PI3K and P53 signaling. Journal of Cellular Physiology. 2018;**233**(6):4634-4642

[99] Laka K, Makgoo L, Mbita Z. Survivin splice variants in arsenic trioxide (As2O3)-induced deactivation of PI3K and MAPK cell Signalling pathways in MCF-7 cells. Genes. 2019;**10**(1):41

[100] Lin L, Liu A, Peng Z, Lin HJ, Li PK, Li C, et al. STAT3 is necessary for proliferation and survival in colon cancer–initiating cells. Cancer Research. 2011;**71**(23):7226-7237

[101] Kroon P, Berry PA, Stower MJ, Rodrigues G, Mann VM, Simms M,

et al. JAK-STAT blockade inhibits tumor initiation and clonogenic recovery of prostate cancer stem-like cells. Cancer Research. 2013;**73**(16):5288-5298

[102] Matsui WH. Cancer stem cell signaling pathways. Medicine. 2016;**95**(1):S8-S19

[103] Ravindran J, Prasad S, Aggarwal BB. Curcumin and cancer cells: How many ways can curry kill tumor cells selectively? The AAPS Journal. 2009;**11**(3):495-510

[104] Dou H, Shen R, Tao J, Huang L, Shi H, Chen H, et al. Curcumin suppresses the colon cancer proliferation by inhibiting Wnt/β-catenin pathways via miR-130a. Frontiers in Pharmacology. 2017;**8**:877

[105] Liu JL, Pan YY, Chen O, Luan Y, Xue X, Zhao JJ, et al. Curcumin inhibits MCF-7 cells by modulating the NF-κB signaling pathway. Oncology Letters. 2017;**14**(5):5581-5584

[106] Tian B, Zhao Y, Liang T, Ye X, Li Z, Yan D, et al. Curcumin inhibits urothelial tumor development by suppressing IGF2 and IGF2-mediated PI3K/AKT/mTOR signaling pathway. Journal of Drug Targeting. 2017;**25**(7):626-636

[107] Shanmugam MK, Rane G, Kanchi MM, Arfuso F, Chinnathambi A, Zayed ME, et al. The multifaceted role of curcumin in cancer prevention and treatment. Molecules. 2015;**20**(2):2728-2769

[108] Yang CL, Liu YY, Ma YG, Xue YX, Liu DG, Ren Y, et al. Curcumin blocks small cell lung cancer cells migration, invasion, angiogenesis, cell cycle and neoplasia through Janus kinase-STAT3 signalling pathway. PLoS One. 2012;7(5):e37960

[109] Chiablaem K, Lirdprapamongkol K, Keeratichamroen S, Surarit R, Svasti J. Curcumin suppresses vasculogenic mimicry capacity of hepatocellular carcinoma cells through STAT3 and PI3K/AKT inhibition. Anticancer Research. 2014;**34**(4):1857-1864

[110] Yamaguchi H, Wyckoff J, Condeelis J. Cell migration in tumors. Current Opinion in Cell Biology. 2005;**17**(5):559-564

[111] Aggarwal BB, Shishodia S, Takada Y, Banerjee S, Newman RA, Bueso-Ramos CE, et al. Curcumin suppresses the paclitaxel-induced nuclear factor-κB pathway in breast cancer cells and inhibits lung metastasis of human breast cancer in nude mice. Clinical Cancer Research. 2005;**11**(20):7490-7498

[112] Bachmeier B, Nerlich A, Iancu C, Cilli M, Schleicher E, Vené R, et al. The chemopreventive polyphenol curcumin prevents hematogenous breast cancer metastases in immunodeficient mice. Cellular Physiology and Biochemistry. 2007;**19**(1-4):137-152

[113] Xia Y, Shen S, Verma IM. NF-κB, an active player in human cancers. Cancer Immunology Research. 2014;**2**(9):823-830

[114] Quintero-Fabián S, Arreola R, Becerril-Villanueva E, Torres-Romero JC, Arana-Argáez V, Lara-Riegos J, et al. Role of matrix metalloproteinases in angiogenesis and cancer. Frontiers in Oncology. 2019;**9**:1370

[115] Sreenivasan S, Thirumalai K, Danda R, Krishnakumar S. Effect of curcumin on miRNA expression in human Y79 retinoblastoma cells. Current Eye Research. 2012;**37**(5):421-428

[116] Yang CH, Yue J, Sims M, Pfeffer LM. The curcumin analog EF24 targets

NF-κB and miRNA-21, and has potent anticancer activity in vitro and in vivo. PLoS One. 2013;**8**(8):e71130

[117] Davoodvandi A, Farshadi M, Zare N, Akhlagh SA, Alipour Nosrani E, Mahjoubin-Tehran M, et al. Antimetastatic effects of curcumin in Oral and gastrointestinal cancers. Frontiers in Pharmacology. 1836;**12**:668567

[118] Tsukamoto M, Kuroda K, Ramamoorthy A, Yasuhara K. Modulation of raft domains in a lipid bilayer by boundary-active curcumin. Chemical Communications. 2014;**50**(26):3427-3430

[119] Montalbán MG, Coburn JM, Lozano-Pérez AA, Cenis JL, Víllora G, Kaplan DL. Production of curcumin-loaded silk fibroin nanoparticles for cancer therapy. Nanomaterials. 2018;**8**(2):126

[120] Krausz AE, Adler BL, Cabral V, Navati M, Doerner J, Charafeddine RA, et al. Curcumin-encapsulated nanoparticles as innovative antimicrobial and wound healing agent. Nanomedicine: Nanotechnology, Biology and Medicine. 2015;**11**(1):195-206

[121] Fan S, Zheng Y, Liu X, Fang W, Chen X, Liao W, et al. Curcumin-loaded PLGA-PEG nanoparticles conjugated with B6 peptide for potential use in Alzheimer's disease. Drug Delivery. 2018;**25**(1):1091-1102

[122] Agarwal NB, Jain S, Nagpal D, Agarwal NK, Mediratta PK, Sharma KK. Liposomal formulation of curcumin attenuates seizures in different experimental models of epilepsy in mice. Fundamental & Clinical Pharmacology. 2013;**27**(2):169-172

[123] Mukherjee A, Sarkar S, Jana S, Swarnakar S, Das N. Neuro-protective role of nanocapsulated curcumin against cerebral ischemia-reperfusion induced oxidative injury. Brain Research. 2019;**1704**:164-173

[124] Taylor RA, Leonard MC. Curcumin for inflammatory bowel disease: A review of human studies. Alternative Medicine Review. 2011;**16**(2):152

[125] Rejinold NS, Thomas RG, Muthiah M, Chennazhi KP, Manzoor K, Park IK, et al. Anti-cancer, pharmacokinetics and tumor localization studies of pH-, RF-and thermo-responsive nanoparticles. International Journal of Biological Macromolecules. 2015;**74**:249-262

[126] Nambiar S, Osei E, Fleck A, Darko J, Mutsaers AJ, Wettig S. Synthesis of curcumin-functionalized gold nanoparticles and cytotoxicity studies in human prostate cancer cell line. Applied Nanoscience. 2018;**8**(3):347-357

[127] Liu R, Pei Q, Shou T, Zhang W, Hu J, Li W. Apoptotic effect of green synthesized gold nanoparticles from curcuma wenyujin extract against human renal cell carcinoma A498 cells. International Journal of Nanomedicine. 2019;**14**:4091

[128] Yallapu MM, Othman SF, Curtis ET, Bauer NA, Chauhan N, Kumar D, et al. Curcumin-loaded magnetic nanoparticles for breast cancer therapeutics and imaging applications. International Journal of Nanomedicine. 2012;**7**:1761

[129] Bhandari R, Gupta P, Dziubla T, Hilt JZ. Single step synthesis, characterization and applications of curcumin functionalized iron oxide magnetic nanoparticles. Materials Science and Engineering: C. 2016;**67**:59-64

[130] Aeineh N, Salehi F, Akrami M, Nemati F, Alipour M, Ghorbani M, et al.

Glutathioneconjugatedpolyethylenimine on the surface of Fe3O4 magnetic nanoparticles as a theranostic agent for targeted and controlled curcumin delivery. Journal of Biomaterials Science, Polymer Edition. 2018;**29**(10):1109-1125

[131] Ayubi M, Karimi M, Abdpour S, Rostamizadeh K, Parsa M, Zamani M, et al. Magnetic nanoparticles decorated with PEGylated curcumin as dual targeted drug delivery: Synthesis, toxicity and biocompatibility study. Materials Science and Engineering: C. 2019;**104**:109810

[132] Bajpai SK, Ahuja S, Chand N, Bajpai M. Nano cellulose dispersed chitosan film with Ag NPs/curcumin: An in vivo study on albino rats for wound dressing. International Journal of Biological Macromolecules. 2017;**104**:1012-1019

[133] Margaritova Zaharieva M, Dimitrov Kroumov A, Dimitrova L, Tsvetkova I, Trochopoulos A, Mihaylov Konstantinov S, et al. Micellar curcumin improves the antibacterial activity of the alkylphosphocholines erufosine and miltefosine against pathogenic Staphyloccocus aureus strains. Biotechnology & Biotechnological Equipment. 2019;**33**(1):38-53

[134] Akhtar F, Rizvi MM, Kar SK. Oral delivery of curcumin bound to chitosan nanoparticles cured plasmodium yoelii infected mice. Biotechnology Advances. 2012;**30**(1):310-320

[135] Karri VV, Kuppusamy G, Talluri SV, Mannemala SS, Kollipara R, Wadhwani AD, et al. Curcumin loaded chitosan nanoparticles impregnated into collagen-alginate scaffolds for diabetic wound healing. International Journal of Biological Macromolecules. 2016;**93**:1519-1529

[136] Yadav VR, Suresh S, Devi K, Yadav S. Novel formulation of solid lipid microparticles of curcumin for anti-angiogenic and anti-inflammatory activity for optimization of therapy of inflammatory bowel disease. Journal of Pharmacy and Pharmacology. 2009;**61**(3):311-321

[137] Kakkar V, Muppu SK, Chopra K, Kaur IP. Curcumin loaded solid lipid nanoparticles: An efficient formulation approach for cerebral ischemic reperfusion injury in rats. European Journal of Pharmaceutics and Biopharmaceutics. 2013;**85**(3):339-345

[138] Nahar PP, Slitt AL, Seeram NP. Anti-inflammatory effects of novel standardized solid lipid curcumin formulations. Journal of Medicinal Food. 2015;**18**(7):786-792

[139] Bhatt H, Rompicharla SV, Komanduri N, Aashma S, Paradkar S, Ghosh B, et al. Development of curcumin-loaded solid lipid nanoparticles utilizing glyceryl monostearate as single lipid using QbD approach: Characterization and evaluation of anticancer activity against human breast cancer cell line. Current Drug Delivery. 2018;**15**(9):1271-1283

[140] Reeves A, Vinogradov SV, Morrissey P, Chernin M, Ahmed MM. Curcumin-encapsulating nanogels as an effective anticancer formulation for intracellular uptake. Molecular and Cellular Pharmacology. 2015;**7**(3):25

[141] Priya P, Raj RM, Vasanthakumar V, Raj V. Curcumin-loaded layer-by-layer folic acid and casein coated carboxymethyl cellulose/casein nanogels for treatment of skin cancer. Arabian Journal of Chemistry. 2020;**13**(1):694-708

[142] Nelson KM, Dahlin JL, Bisson J, Graham J, Pauli GF, Walters MA. The

essential medicinal chemistry of curcumin: Miniperspective. Journal of Medicinal Chemistry. 2017;**60**(5):1620-1637

[143] Basniwal RK, Buttar HS, Jain VK, Jain N. Curcumin nanoparticles: Preparation, characterization, and antimicrobial study. Journal of Agricultural and Food Chemistry. 2011;**59**(5):2056-2061

[144] Adahoun MA, Al-Akhras MA, Jaafar MS, Bououdina M. Enhanced anti-cancer and antimicrobial activities of curcumin nanoparticles. Artificial Cells, Nanomedicine, and Biotechnology. 2017;**45**(1):98-107

[145] Rai M, Pandit R, Gaikwad S, Yadav A, Gade A. Potential applications of curcumin and curcumin nanoparticles: From traditional therapeutics to modern nanomedicine. Nanotechnology Reviews. 2015;**4**(2):161-172

[146] Yavarpour-Bali H, Ghasemi-Kasman M, Pirzadeh M. Curcumin-loaded nanoparticles: A novel therapeutic strategy in treatment of central nervous system disorders. International Journal of Nanomedicine. 2019;**14**:4449

[147] Gupta V, Aseh A, Ríos CN, Aggarwal BB, Mathur AB. Fabrication and characterization of silk fibroin-derived curcumin nanoparticles for cancer therapy. International Journal of Nanomedicine. 2009;**4**:115

[148] Teixeira CC, Mendonça LM, Bergamaschi MM, Queiroz RH, Souza GE, Antunes LM, et al. Microparticles containing curcumin solid dispersion: Stability, bioavailability and anti-inflammatory activity. AAPS PharmSciTech. 2016;**17**(2):252-261

[149] De Leo V, Milano F, Mancini E, Comparelli R, Giotta L, Nacci A, et al. Encapsulation of curcumin-loaded liposomes for colonic drug delivery in a pH-responsive polymer cluster using a pH-driven and organic solvent-free process. Molecules. 2018;**23**(4):739

[150] Vijayakurup V, Thulasidasan AT, Retnakumari AP, Nandan CD, Somaraj J, Antony J, et al. Chitosan encapsulation enhances the bioavailability and tissue retention of curcumin and improves its efficacy in preventing B [a] P-induced lung carcinogenesis. Cancer Prevention Research. 2019;**12**(4):225-236

[151] Gera M, Sharma N, Ghosh M, Huynh DL, Lee SJ, Min T, et al. Nanoformulations of curcumin: An emerging paradigm for improved remedial application. Oncotarget. 2017;**8**(39):66680

[152] Peng S, Li Z, Zou L, Liu W, Liu C, McClements DJ. Improving curcumin solubility and bioavailability by encapsulation in saponin-coated curcumin nanoparticles prepared using a simple pH-driven loading method. Food & Function. 2018;**9**(3):1829-1839

[153] Cheng C, Peng S, Li Z, Zou L, Liu W, Liu C. Improved bioavailability of curcumin in liposomes prepared using a pH-driven, organic solvent-free, easily scalable process. RSC Advances. 2017;**7**(42):25978-25986

[154] Chabib L, Martien R, Ismail H. Formulation of nanocurcumin using low viscosity chitosan polymer and its cellular uptake study into T47D cells. Indonesian Journal of Pharmacy. 2012;**23**(1):27-35

[155] Bisht S, Feldmann G, Soni S, Ravi R, Karikar C, Maitra A, et al. Polymeric nanoparticle-encapsulated curcumin (" nanocurcumin"): A novel strategy for human cancer therapy. Journal of Nanobiotechnology. 2007;**5**(1):1-8

[156] Sandhiutami NM, Arozal W, Louisa M, Rahmat D, Wuyung PE. Curcumin nanoparticle enhances the anticancer effect of cisplatin by inhibiting PI3K/AKT and JAK/STAT3 pathway in rat ovarian carcinoma induced by DMBA. Frontiers in Pharmacology. 2021;**11**:2199

[157] Wu Q, Ou H, Shang Y, Zhang X, Wu J, Fan F. Nanoscale formulations: Incorporating curcumin into combination strategies for the treatment of lung cancer. Drug design, development and. Therapy. 2021;**15**:2695

[158] Mohan Yallapu M, Ray Dobberpuhl M, Michele Maher D, Jaggi M, Chand CS. Design of curcumin loaded cellulose nanoparticles for prostate cancer. Current Drug Metabolism. 2012;**13**(1):120-128

[159] Chirio D, Gallarate M, Peira E, Battaglia L, Serpe L, Trotta M. Formulation of curcumin-loaded solid lipid nanoparticles produced by fatty acids coacervation technique. Journal of Microencapsulation. 2011;**28**(6):537-548

[160] Lim KJ, Bisht S, Bar EE, Maitra A, Eberhart CG. A polymeric nanoparticle formulation of curcumin inhibits growth, clonogenicity and stem-like fraction in malignant brain tumors. Cancer Biology & Therapy. 2011;**11**(5):464-473

[161] Singh SP, Sharma M, Gupta PK. Enhancement of phototoxicity of curcumin in human oral cancer cells using silica nanoparticles as delivery vehicle. Lasers in Medical Science. 2014;**29**(2):645-652

[162] Yallapu MM, Ebeling MC, Jaggi M, Chauhan SC. Plasma proteins interaction with curcumin nanoparticles: Implications in cancer therapeutics. Current Drug Metabolism. 2013;**14**(4):504-515

[163] Sheikh E, Bhatt ML, Tripathi M. Role of nano-curcumin: A treatment for cancer. The Journal of Medicinal Plants. 2017;**5**:394-397

[164] Mimeault M, Batra SK. Potential applications of curcumin and its novel synthetic analogs and nanotechnology-based formulations in cancer prevention and therapy. Chinese Medicine. 2011;**6**(1):1-9

[165] Tabanelli R, Brogi S, Calderone V. Improving curcumin bioavailability: Current strategies and future perspectives. Pharmaceutics. 2021;**13**(10):1715

[166] Zielińska A, Costa B, Ferreira MV, Miguéis D, Louros J, Durazzo A, et al. Nanotoxicology and nanosafety: Safety-by-design and testing at a glance. International Journal of Environmental Research and Public Health. 2020;**17**(13):4657

[167] Sanna V, Siddiqui IA, Sechi M, Mukhtar H. Nanoformulation of natural products for prevention and therapy of prostate cancer. Cancer Letters. 2013;**334**(1):142-151

[168] Vega-Villa KR, Takemoto JK, Yáñez JA, Remsberg CM, Forrest ML, Davies NM. Clinical toxicities of nanocarrier systems. Advanced Drug Delivery Reviews. 2008;**60**(8):929-938

[169] Długosz O, Szostak K, Staroń A, Pulit-Prociak J, Banach M. Methods for reducing the toxicity of metal and metal oxide NPs as biomedicine. Materials. 2020;**13**(2):279

Section 2

Production Technology

Chapter 8

Sustainable Ginger Production through Integrated Nutrient Management

N. Divyashree, S. Poojashree, S. Venukumar and Y.C. Vishwanath

Abstract

The spice ginger is one of the most extensively used species in the Zingiberaceae family. It is frequently used as a condiment with many different cuisines and drinks. In addition to being used as a spice, it is a key component in both conventional and modern medicine. It strengthens immunity and is a rich source of several minerals and physiologically active compounds. Since it can be grown in a variety of climatic circumstances, the production of this spice has been increasing in most regions of the world. Because it is a nutrient-exhaustive crop that needs an appropriate supply of nutrients at critical stages of its growth in the form of chemical fertilisers or organic manuring, or a combination of both. To obtain excellent quality and quantity of ginger rhizomes as well as protect soil health and environmental quality, effective nutrient management can aid in decreasing the abuse of chemical fertilisers. In this perspective, this chapter aims to depict Integrated Nutrient Management (INM) for the sustainable production of ginger, as INM is a crucial component of sustainable agriculture, which necessitates resource management in a way to satisfy changing human requirements without degrading the quality of the environment and conserving essential natural resources.

Keywords: ginger, organic nutrient management, inorganic and integrated nutrient methods, nutrient uptake, nutrient use efficiency, organic farming

1. Introduction

Ginger is the root of the plant *Zingiber officinale* roscoe, which is a member of the Zingiberaceae family. It is among the most widely used spices and healing agents in the world. The plant is referred to as Sringavera in Sanskrit, and it is possible that this term evolved into Zingiberi in Greek and then Zingiber in Latin [1]. It is mostly employed in food as a spice and flavouring agent. It is widely used as an ingredient in gingerale, candies, pastries, and cakes in the food business [2]. In cookery, ginger is used in a variety of forms, including young ginger, mature fresh ginger, dry ginger, ginger oil, ginger oleoresin, dry-soluble ginger, paste, and ginger emulsion [3]. It has a lot of oleoresins, which are secondary metabolites and are important for flavour and pungency.

Numerous studies have been conducted to learn more about this plant's miraculous properties. It is a vital tropical agricultural crop and a significant source of spices, though its benefits are not just limited to cooking, but also in different pharmaceutical preparations [4], as it is a great source of several bioactive phenolics, including non-volatile pungent substances including gingerols, paradols, shogaols, and zingerones. Since antiquity (more than 2500 years) ginger has also been used in traditional oriental medicine (Ayurvedic, Chinese, and Unani systems of medicine) to treat a variety of illnesses, such as rheumatoid arthritis, sprains and muscular aches, sore throats, nausea, constipation, indigestion, fever, infectious diseases, and helminthiasis [5, 6].

Ginger is a versatile home remedy that can be used to treat a variety of conditions, including motion sickness, arthritis, diarrhoea, the flu, headache, heart, and menstruation difficulties. Numerous research has been conducted on ginger's ability to treat complex illnesses like cancer and chronic migraines. Here are some of the ginger's active ingredients that have a variety of medicinal uses (**Table 1**).

Ginger has wide range of applications in food and pharmaceutical industries. Hence, there is a greater demand at global level. So, it is essential to increase the production rate by supplementing with balanced nutrients for their better growth and development.

Active compound of ginger	Biological activities	References
Gingerol and gingerol related compound	The antioxidant activity.	[7]
	Anti-tumour activity via induction of apoptosis, modulation of genetic and other biological activity.	[8, 9]
	Anti-inflammatory and anti-analgesic activity.	[10]
	Anti-microbial activity.	[11]
	Hepato-protective activity.	[12, 13]
Paradol	Antioxidant and anti-cancerous activity.	[14, 15]
	Antimicrobial activity.	[16]
Shogoal	Antioxidant and anti-inflammatory activity.	[8]
	Shogaol showed anticancer activities through the inhibition of cell invasion reduction of matrix metalloproteinase-9 expression, anti-proliferation activity and anti-invasion.	[17–19]
Zingerone	Antioxidant activity.	[20, 21]
	Anti-inflammatory action.	[22, 23]
	Anti-bacterial activity.	[24]
Zerumbone	Anti-tumour activity.	[25]
	Anti-microbial activity.	[26]
1-Dehydro-(10) gingerdione	Regulation of inflammatory genes.	[27]
Terpenoids	Induce Apoptosis by activation of p53.	[28]
Ginger flavonoids	Antioxidant activity.	[29]

Table 1.
Biological activities of ginger active compounds.

2. The significance of the nutrients in ginger

The availability of nutrients, which is controlled by their distribution and rates of cycling in the soil, has an impact on crop primary production. Nutrient components play key roles in the physiological activities of plants. The two key nutrients that are most deficient in Indian soils are nitrogen (N) and phosphorus (P). Deficits in potassium (K) and sulphur (S) can arise in particular regions and soil types. Acidic soils lack calcium (Ca) and magnesium (Mg), which must be supplemented for excellent agricultural yields. Micronutrient deficiencies of zinc (Zn), copper (Cu), iron (Fe), manganese (Mn), boron (b), molybdenum (Mo), and sulphur (S) have been discovered to be widespread in Indian soils [30]. In order to sustain soil fertility, it is crucial to replenish the minerals and sequester organic C. The information in this chapter covers a wide range of topics related to managing nutrients for ginger crops. It should be noted that managing nutrients based on soil testing is an effective management tool that should be used to make fertiliser recommendations and maintaining soil quality based on site-specific nutrient management techniques should be one of our top priorities.

Nutrient management is essential in achieving the best growth and productivity in ginger, in addition to soil type. Because it is a crop that exhausts nutrients, it needs a sufficient supply during key growth periods [31]. Chemical fertilisation (also known as "chemical nutrient management" or "CNM"), organic manuring (also known as "organic nutrient management" or "ONM"), or a combination of both (also known as "integrated nutrient management" or "INM") are all alternatives for managing the crop's nutrients [32].

However, reports indicate that depending on the crop variety, soil type, and geographic region, different amounts of nutrients may be needed, whether through organic or chemical methods. The recommended dose of fertilisers (RDF) should ideally be applied in splits to satisfy the crop's requirement at different phases of growth, and a good nutrient management plan can significantly reduce the discrepancy between prospective yield and actual output. The usage of chemical fertilisers can be reduced as a result, protecting the environment's quality [33–37].

3. Fertilisers recommendations for ginger

Ginger is an exhausting crop and benefits greatly from the application of nutrients at particular dosage. The need for nutrients varies according on the soil type, crop growth stage, variety and location. The NPK values for ginger have so been standardised by certain experimental trials.

For the entire nation, the AICRP (All India Coordinated Research Project) on spices has recommended 100 kg of nitrogen and 50 kg each of phosphorus and potassium. Application of 1/3 N, full P_2O_5, half or full K_2O, and 1/3 N 1 month or 40–60 days after planting are recommended. The final 1/3 N should be applied 2–3 months after planting [3]. According to ref. [38], the application of 60 kg N, 40 kg P_2O_5, and 60 kg K_2O/ha produced good results. According to ref. [39], the effect of N and P on ginger output was inconsequential. Ref. [40] reported that N at 50–100 kg/ha had significantly increased the yield of ginger by 18 to 32 per cent. For a superior yield of ginger under Kerala conditions, ref. [41] advised using 60 N, 60 P_2O_5, and

150 K kg/ha. The optimal application was 100 kg of N, 50 kg of $P_2 0_5$, and 50 kg of $K_2 0$ per hectare, according to ref. [42]. According to ref. [43], the highest yield of 43 tonnes of green ginger per hectare was achieved with 100 kg N, 100 kg $P_2 0_5$, and 200 kg $K_2 0$ per hectare. The need of N application for ginger at the active development stage, or 120–135 days after planting, and the tillering stage, or 200–210 days after planting, was emphasised by ref. [44]. With NPK at 60:60:120 kg/ha, ref. noticed a modest increase in yield. They also stated that when $K_2 0$ was increased from 80 to 120 kg/ha, the yield of ginger increased. The NPK dose of 80:30:40 Kg per acre was found to be ideal in studies done at Kerala Agricultural University Vellanikkara. It was discovered that the combination application of N and K had a stronger effect than each compound acting alone. N and K recorded the maximum yield at N-180 K–160 kg/ha [45].

3.1 Drawbacks of fertilisers use alone

Inorganic fertiliser usage typically necessitates more frequent fertilising. Plants can easily take the nutrients, but they do not stay in the soil for as long. This issue can be addressed by using a slow-release fertiliser, but typically, inorganic fertiliser needs to be applied to the soil often. Although inorganic fertilisers are frequently less expensive, the cost savings may be offset by the need for several applications because it depends on the amount of nutrients in soil. Plants can be burned or scorched if too much fertiliser is used because the nutrients are concentrated and easily accessible. High salinity is also more likely when using inorganic fertilisers. Saline soils contain an excessive amount of salt and can impair a plant's ability to absorb nutrients and water. The soil's surface may develop a crust as a result of these fertilisers.

4. Influence of biofertilizers on nutrition in the production of ginger

To restore the soil's fertility, biofertilizers are necessary. Use of chemical fertilisers over a lengthy period damages the soil and reduces crop output. On the other side, biofertilizers improve the soil's ability to hold water while also adding vital minerals like nitrogen, vitamins, and proteins. Since they are a natural source of fertiliser, agriculture uses them extensively. Crop yields are said to be increased by bioinoculants like *Arbuscular Mycorrhizal Fungi* (AMF) and Trichoderma, which have growth-promoting properties. They are said to enhance root growth, which results in improved crop nutrient uptake and higher yields. With the application of these bio inoculants, tolerance to abiotic stress and resistance against plant diseases have also been reported [46]. Additionally, AMF inoculation aids in the selective activation of advantageous soil bacteria [47]. The members of the Glomeraceae family provided greater protection from pathogen incursions, while the AMF belonging to the Gigasporaceae family showed noteworthy nutrient absorbing capacity [48]. According to ref. [49], the use of IISR Power Mix G @ 0.5% at 2 and 3 months after planting increased the production of ginger by 11% compared to control. Higher fresh ginger production was achieved in Kerala by using the microbial inoculants AMF (5 g/plant) and Trichoderma (1 kg combined with 90 kg FYM and 10 kg neem cake and 250 g of the combination utilised) during planting time and Trichoderma 1 month after planting. In Himachal Pradesh,

India, Azospirillum inoculation with VAM (*Vesicular-Arbuscular Mycorrhizal*) and *Glomus mosseae* improved ginger plant growth and yield [50].

5. Foliar sprays' nutritional impact on ginger production

A foliar application is when liquid fertiliser is sprayed on top of leaves rather than the soil to feed plants. Through their stomata and epidermis, the absorption happens. Although total absorption may be just as great through the epidermis, transport is typically faster through the stomata. Most crops experience periods of elevated nutrient need during regular seasonal development. Typically, this increase involves a wide range of macro, meso, and micro factors. The following important growth phases of ginger's life cycle require an immediate supply of nutrients:

- Rapid seedling development following germination
- Tillering
- Vigorous clump growth
- Increased requirement for B and Ca during flowering, which is necessary for the growth and development of pollen tubes and
- Rhizome development

During these critical growth stages, the rapid nutrient supply to the plants is essential, where this cannot be achieved rapidly through soil application, where foliar spray can.

There are reports of increased ginger growth and yield when $ZnSO_4$ (0.5%) spray is used [51]. Foliar fertilisers with moderate release rates can therefore effectively offset the decreased Zn bioavailability and mobility [52]. In comparison to soil application of higher levels of fertilisers, foliar application of lower levels of fertilisers, particularly urea, results in higher yields for several crops. Additionally, it aids in raising the produce's quality. In 2021, ref. [53] reported on the usage of growth substances like NAA to enhance growth and development as well as yield. The performance of enhanced varieties like Rio-de Janeiro and Baruwa Sagar with single and mixed application of urea and NAA was evaluated in a field trial under the aforementioned conditions. The yield per plant demonstrated the variety, urea, and NAA primary effects, as well as the interaction impact of urea and NAA. Baruwa Sagar's diversity was vastly superior. The combination of NAA 400 PPM and urea 2% produced the highest rhizome production per plant. Higher yield was produced using urea at 2% and NAA at 400 PPM. However, best response was reported in case of urea 2%. Spraying urea (2.0%) and planofix (200 ppm) was proven to increase the output of dry ginger. The seed rhizomes were exposed to 250 ppm ethephon for 15 minutes, and this resulted in a noticeable increase in shoot and root growth at the early stage. When applied as a foliar spray every 15 days beginning at 70 DAP, ethrel (200 ppm) increases the quantity of tillers and leaves. We can use CCC (50–125 ppm) to prevent shoot growth while it's being stored. According to one assay, ginger plants have very little gibberellin, and their rhizomes contain cytokinins and auxins that have an impact on the start and growth of rhizomes. Auxin and cytokinin levels are raised by CCC spray. Shadap [54] based on growth, yield and net returns and benefit: cost ratio point of view, reported that spraying with Zn 0.5%. followed by B 0.3% is best to get maximum yield. Clumps supplied with 100 per cent RDF (NPK) as soil application along with BA spray at 75 ppm recorded maximum growth and yield in transplanted ginger at Mudigere, Karnataka [55]. Supplemental foliar nourishment is a booster to attain better yield by correcting soil limitations (51).

6. Cultural practises and its nutrients influence on ginger cultivation

Mulching the ginger beds with green leaves is an essential operation to enhance germination of seed rhizomes and to prevent washing off soil due to heavy rain.

Several employees from various sections of the country have cited the advantages of mulching ginger. A total of 20 tonnes of green leaves were used as leaf mulch during planting and 6 weeks later, yielding a crop that was 200% more productive than one that wasn't mulched [41]. Under Wynad circumstances, 15 tonnes of green leaves were sufficient for mulching [40]. Mulching was deemed necessary by ref. [42] at rates of 2.5, 5.0, and 5.0 t/ha for the first, second, and third mulchings, respectively. Heavy mulch, according to ref. [45], may alter the soil's physical and chemical environment, increasing the availability of phosphorus and potassium. In Bihar, it was discovered that mulching with shisan leaves was superior to mulching with paddy straw, mango leaves, or neither. Shisan acted as an organic manure to increase the output . Mulching with dry sal leaves was advised by the CPCRI at Kasaragod, Kerala State, India. With a treatment of FYM at 30 t/ha, intercropping ginger under *Ceiba pentandra* produced a greater yield and revenue than the main crop did when it was 25 per cent trimmed. Since ginger is a crop that prefers shade, it produces well with the addition of organic matter when planted in the best shade [56]. Paddy straw and *Schima wallichii* dry leaf mulches, which are both locally accessible organic mulches, enhanced ginger yield in Meghalaya by 43.6 and 39.7%, respectively, when applied at a rate of 16 t/ha. Both using live soybean intercrops as mulch and mulching three times with leaves were found to be equally efficient. The recommended mulching rates for ginger cultivation are 12.5, 5.0, and 5.0 t/ha for the first, second, and third mulchings, respectively (**Table 2**) [42].

7. Nutritional influence of organic nutrients on ginger cultivation

The use of organic manures held a very prestigious position with farmers in the past but subsequently, the introduction of high analysis chemical fertilisers cast a shadow on their spread [57]. It is well known that addition of organic manures has shown considerable increase in crop yield, quality and exert significant influence on physical, chemical and biological properties of soil. Use of organic and biofertilizers not only improve soil health but also help to sustain crop productivity.

Ginger must be grown organically in areas that are 25 metres wide apart from conventional farms on all sides. This isolation belt's produce must not be treated as organic. Being an annual crop, a two-year conversion period will be needed. Ginger can be produced organically as a companion or mixed crop as long as all the other plants are also grown that way. It is preferable to rotate ginger with a leguminous crop. You can adopt ginger-banana-legume or ginger-vegetable-legume (recommended

Mulch	Yield (Q/ha)
Shisan leaves	78.25
Mango leaves	72.76
Paddy straw	52.39
No mulch	28.25

Table 2.
Effect of mulching on ginger.

by Tamilnadu Agricultural University). When planted with coconut, arecanut, mango, leucaena, young rubber plantations, etc., recycling of farm waste can be done successfully. Ginger can be the best component crop in agri-horticulture and silvi-horticulture systems. It can be grown or rotated as a mixed crop with green manure, legumes, or trap crops to effectively build up nutrients and manage pests and diseases. Every crop in the field must be treated to organic production practises when cultivated in a mixed farming system. For the organic production of ginger cultivated as an intercrop in coconut gardens, several organic manures including FYM, vermicompost, neem cake, and green leaf manures, as well as microbial inoculants including AMF and Trichoderma and their mixtures, were tested. FYM (30 t/ ha) + neem cake + AMF + Trichoderma and FYM + AMF both generated noticeably greater yields than other treatments among the many combinations tested [58].

Ginger needs organic matter, which can be obtained from a variety of sources, including mulches and green/organic manures. This was well demonstrated by the successful crop production in the high fertility conditions of Wayanad, Kerala State, India, which received 15 t of green leaf mulch and 10 t of organic manure per hectare, all without the use of chemical fertilisers [39]. When humus and organic matter are available, ginger grows well and has a favourable relationship with yield [59]. Most of the organic manures are applied in basal doses, while they are occasionally used as mulch after a crop has emerged in some regions. However, farmers in Maharashtra use a lot of FYM—40–50 t/ha on average. The recommended amount of organic manure for Kerala is 30 t/ha of green leaf mulch and 25–30 t/ha of FYM applied in three separate applications. Kerala had the highest yield and benefit-cost ratio from applying FYM up to 48 t/ha [60]. The AICRP conducted field tests on spices at several sites in India, and the results showed that the use of multiple organic sources, including FYM, pongamia oil cake, neem oil cake, stera meal, rock phosphate, and wood ash, produced results comparable to those of the standard method. Farmyard manure may be applied at a rate of 25–30 t/ha, together with vermicompost at a rate of 4 t/ha and green leaf mulching at a rate of 12–15 t/ha every 45 days. The fertility and production will also be increased by adding oil cakes like neem cake (2 t/ha), composted coir pith (5 t/ha), and suitable microbial cultures of Azospirillum and phosphate-solubilising bacteria. To promote growth and control disease, the use of the PGPR strain of *Bacillus amyloliquefaciens* (GRB 35) is also advised. To obtain the necessary amount of phosphorus and potassium, it may be necessary to apply lime/dolomite, rock phosphate, and wood ash depending on the results of the soil test. For a higher production within the parameters of standard setup of certifying organisations, restricted use of foliar spraying of micronutrient mixture specifically for ginger is recommended (dosage @ 5 g/L) twice, at 60 and 90 DAP. Ref. [61] reported highest yield/ha (32.88 t/ha) was noticed in poultry manure @ 2 t/ha followed by Mustard cake @ 0.75 t/ha. The highest curcumin % and oleoresin % were noticed in

Vermicompost @ 5 t/ha over control in the Gangetic Alluvial Plains of West Bengal. Ref. [62] reported that, organic manures in the forms of cow dung, poultry and pig manures have great tendency to increase growth characters and yield of ginger in a rainforest zone, Nigeria.

The primary approach used in an organic system for managing insect pests and diseases is the use of biocontrol agents in conjunction with cultural and phytosanitary measures. The shoot borer can be controlled with an integrated strategy that includes trimming and removing newly infested shoots from July through August (at fortnightly intervals) and spraying Neemgold 0.5% or neem oil 0.5% from September through October (at 21-day intervals). To control the rhizome rot disease, it is possible to choose healthy rhizomes, solarize the soil and add Trichoderma, treat the seeds, and then apply biocontrol agents like Trichoderma, PGPR, or Pseudomonas multiplied in suitable carrier media like coir pith compost, well-rotten cow dung, or high-quality neem cake to the soil. Bordeaux mixture 1 per cent may be sprayed to control various foliar diseases, but only in an amount not to exceed 8 kg of copper per hectare per year. To control the nematode population, apply the high-quality neem cake indicated earlier and the bioagent *Pochonia chlamydosporia.*

Trials on different management systems on ginger at the IISR, Calicut, Kerala State, India, showed that higher soil nutrient build-up with the highest organic carbon content (2.33%) was in the organic system, which was on par with the integrated system of nutrient management and among the different systems of nutrient management. The maximum soil P, Ca, Mg, Zn, and Cu availability was found in the organic management method. The impact of various cropping systems on the microbial community in soil also revealed that the organic nutrient management system had the highest concentrations of *Pseudomonas fluorescens, Azospirillum,* and *phosphobacteria*. The activities of enzymes, such as dehydrogenase, acid phosphatase, alkaline phosphatase, cellulase, and urease, were significantly higher under the organic system of nutrient management as compared to the exclusive inorganic system or integrated system of nutrient management. However, during the initial years, 15–20% reduction in yield under the organic system of nutrient management was encountered [63].

There is lot of scope to popularise the organic ginger produce for export in foreign country. Commercially the ginger is produced at north eastern states organically by default because the farmers of the region neither apply the chemical fertilisers nor chemical pesticides in ginger crop. They are only applying the locally available farmyard manures (cow dung, pig manure, poultry manures, rabbit manure, *etc.,*) in whole north-eastern region. In this way, the ignorance of the farmers about the technological advances is turning out to be a key to prosperity. Considering the increasing demand for organic produce all over the world, the farmers can hope to get better returns for their produce [64].

7.1 Limitations in use of organic manures alone

- Limited effectiveness of organic fertilisers as they release nutrients into the soil with the help of microorganisms that break down the fertilisers slowly. And for the microorganisms to break down the organic fertilisers effectively, they need warmth and moisture. If the soil is not warm or moist enough, the breakdown of the organic fertiliser will take time and its effects will be limited.

- Slow breakdown of the organic fertilisers in the soil
- Organic fertiliser is more expensive than chemical fertiliser

8. Integrated nutrient management (INM) in ginger

It is crucial to provide the soil with sufficient levels of vital nutrients in a balanced proportion at the proper time and in the right manner for the cultivation of any crop. The combined effect of organic and inorganic fertilisers in integrated nutrient management strategies would be a useful approach to acquire higher yield and higher-quality produce, as described above, both of which have their own downsides. Every crop should use integrated nutrient management strategies, which combine different inorganic, organic, and biological sources of nutrients. The methods used in conventional and organic production differ fundamentally [65, 66]. In addition to the biomass yield of a crop, the nutrient management techniques used for its cultivation are likely to have an impact on the crop's quality, the soil's fertility, and the overall economics of its cultivation. Talk about the modern idea of "farming for health," the sustainability of natural resources, notably soil, and lastly the stability of the farming community's financial situation. Numerous investigations on INM in ginger have been published.

The Kerala Agriculture University (KAU) has suggested a general nutrient dosage for ginger for the state of Kerala. Well-decomposed cattle manure or compost must be applied at the time of planting, either by disseminating it over the beds prior to planting or applying it in the planting pits. Neem cake, applied at a rate of 2 tonnes per hectare during planting, aids in decreasing the prevalence of rhizome rot disease/nematode and improving production. Two to three separate applications of the fertilisers are required. At the time of planting, a full amount of phosphorus is sprayed as basal. At 45, 90, and 120, equal split doses of N and K are top dressed (120 DAP. A basal application of zinc fertiliser up to 6 kg/ha (30 kg of zinc sulphate per hectare) in soils lacking in zinc produces good yield. For a greater yield, foliar treatment of a micronutrient mixture tailored for ginger is also advised (dosage @5 g/L) twice, at 60 and 90 DAP.

By considering different soil features, plant production and quality, crop and soil quality, and economics, ref. [67] thoroughly assessed the various nutrient management strategies used in ginger from 2015 to 2019 in Punjab. Farmyard manure (FYM), vermicompost (VC), urea, single superphosphate, and muriate of potash were among the 14 nutrient management techniques used. According to the study's findings, applying 75% RDN plus 25% N through VC resulted in higher crop yield and crop quality, with yields rising by 103.1%, 21.9%, and 75.7%, respectively, above absolute control, RDN, and organic management. Under this treatment, the maximum harvest index (71.4%) and crop quality index (20.0) were obtained. According to reports, 25 to 30 tonnes of bovine dung and an 8:8:16 fertiliser mixture applied at 450 kg/ha were helpful in Kerala to increase production.

Ref. [57] conducted a field experiment to know response of organic manures and fertilisers to yield and nutrient uptake of ginger (*Z. officinale* Rosc.)" was conducted at Agronomy Farm, College of Agriculture, Pune during summer 2006. The data revealed that recommended dose of fertiliser +25 t FYM/ha favourably influenced yield and uptake of nutrients by ginger followed by the application of 50% N through recommended dose +50% N through poultry manure. It is, therefore suggested that application of recommended dose of fertiliser +25 t FYM/ha to ginger planted on flat bed in clay loam soil is best combination.

The highest fresh rhizome yield (1.87 t/ha), lowest rhizome rot (11%) and oleoresin content (5.82%), were obtained with 100% recommended rates of fertilisers, along with Azospirillum application at a rate of 10 kg/ha combined with FYM at 10 t/ha [68]. Soil application of Gigaspora at the time of planting (2.5 g/ rhizome) was also found to increase the yield as in the case of pine needle organic amendment and seed treatment with *T. harzianum*. Also, the effects of humic acid fertiliser on soil urease activity and available N content, N uptake, and rhizome yield were reported [69]. Ref. [69] reported that the highest yield of rhizome in tribal areas of Orissa as obtained with farm yard manure at 25 t + NPK 75, 50 and 50 kg/ha respectively. The yield of ginger was more when 20 t of FYM and 125 months gave maximum green ginger yield [70].

In ginger, oil content did not vary significantly among the treatments, as shown in **Table 3**. However, the fibre content was significantly reduced in the organic system of nutrient management. Interestingly, both oleoresin and starch contents were the maximum in the organic system of nutrient management, and, in both cases, there were statistically significant differences among the three systems of nutrient management. The maximum yield and oleoresin content was obtained with the application of 10 t/ha of FYM + 1.25 t/ha of compost +20 kg/ha of Azospirillum, which also showed higher nutrient uptake [71].

8.1 Soil quality under INM in ginger

It is critical to investigate how integrated nutrient management regimes affect the biochemical and microbiological characteristics of soils used for ginger growth [72]. There are, however, very few papers that examine the effects of various nutrition regimens on ginger yield and quality while also involving a number of field tests. However, it is crucial to simultaneously determine how they affect a variety of soil physicochemical and biological properties [73].

Organic and integrated nutrient management resulted in a decrease in bulk density and consequent increase in soil porosity. While subsequent modifications brought about by organic nutrient management were in the scale of a 9.2% drop in bulk density and an 11.6% increase in porosity, integrated nutrient management led to a 5.7% decrease in bulk density and a 7.0% increase in porosity. These findings are supported by the fact that the decomposition of organic materials generated organic acids, which directly altered soil pH and indirectly affected bulk density by forming soil aggregates and increasing soil porosity.

By taking a variety of soil conditions into account, ref. [66] thoroughly assessed the various nutrient management strategies used in ginger from 2015 to 2019 in

Management system	Oil content (%)	Oleoresin content (%)	Starch (%)	Fibre (%)
Organic	1.20	3.96	70.07	1.69
Inorganic	1.20	3.15	62.21	1.90
Integrated	1.25	3.36	55.82	1.90
LSD (95%)	NS	0.23	9.89	0.08

Table 3.
Effect of different nutrient management systems on the quality of ginger.

Punjab. The study found that the highest soil quality index (SQI) was achieved with the 100% recommended dose of nitrogen (RDN) and FYM, whereas 100% NPK through FYM increased the soil's organic carbon, physical qualities, and microbiological characteristics. SQI grew to 0.63 with integrated nutrient management and to 0.36 with organic management.

Cropping requires ploughing, which upset the stability and distribution of soil aggregates [74], exposing soil organic C to quick oxidation. So, when ginger was grown without organic supplementation, soil organic carbon decreased by 4.1% under control nutrient management, whereas SOC (Soil Organic Carbon) increased by 24.3% under integrated nutrient management. Also, plant residue C would have been deposited as a result of INM because of its favourable effects on root, vegetative growth, and yield. The results showed that FYM was more effective at raising SOC than VC among INM treatments (Vermicompost). This is because FYM has more lignin, polyphenols, and a greater C/N ratio than VC. Thus, the FYM-C was more resistant to breakdown than the VC due to greater lignin and polyphenol concentrations that resulted in the formation of stable complexes with proteins of plant origin. Due to this fact, FYM-treated plots outperformed VC-treated plots in terms of maximal breathing capacity (MBC). MBC and microbial activity in these soils increased due to the application of organics, either alone or in combination with inorganics, which created a more suitable environment for rapid microbial growth. The direct addition of nutrients through organic manures and enhanced activity of soil microorganisms, which converted organically bound nutrients to inorganic/available forms in the soil, may be blamed for an increase in available nutrients through INM over organics and control [68]. By momentarily immobilising the chemical fertiliser, the organic manures would have also improved its effectiveness by lowering the leaching of plant nutrients. The solubilisation effect of organic acids generated from the breakdown of organic manures on applied SSP (20% Ca) and native soil Ca may also be responsible for the increased exchangeable Ca under INM treatments compared to control. All micronutrient cation (Mn, Cu, and Zn) contents were strongly impacted by the combination of organic and inorganic sources, with the exception of Fe [37]. The creation of higher solubility organic chelates and mineralisation of organically bound forms, which reduces the susceptibility of the micronutrients to adsorption, fixation, and/or precipitation, could also be the cause of this rise in micronutrients in comparison to the control treatment, the application of cow dung and poultry litter enhanced the soil's pH, organic matter, total nitrogen content, accessible P and K contents, and exchangeable K, Ca, and Mg contents [75].

8.2 Nutrient removal, absorption, usage effectiveness, and indices of INM in ginger

Rhizomes of ginger primarily remove N and K, remove P and Mg to a lesser extent, and remove Ca to the least extent [76]. According to ref. [77], a buildup of macronutrients in the decreasing sequence of N, K, Ca, Mg, S, and P, as well as micronutrients in the order of Fe, Mn, Zn, B, and Cu. However, nutrient uptake varies greatly depending on the kind of soil, the climate, the amount of nutrients in the soil, and the variety or cultivar grown. Three unique growth phases can be used to categorise ginger's development: active growth (90–120 DAP), sluggish vegetative growth (120–180 DAP), and senescence (180 DAP), during which the rhizome continues to develop up to harvest. According to ref. [78], ginger shoots and leaves are the areas where the majority of the assimilated carbon (C) is transported at the seedling stage. Following that, as the plant grew, the distribution rate into the rhizome gradually dropped while it steadily increased for shoots and leaves. The rhizome becomes the growth centre during the rhizome's stage of rapid growth because C is mostly transferred from the leaves to the rhizomes at this time. N was absorbed and used in the same ways that C assimilates were. At seedling stage, the shoots and leaves received around 48.41% of the nitrogen (N) absorbed from the fertiliser applied. While 65.43% of the N came from fertiliser applied at different. At seedling stage, the shoots and leaves received around 48.41% of the nitrogen (N) absorbed from the fertiliser applied. While 65.43% of the N from the fertiliser provided at different phases of the rhizomes' growth went to the rhizomes, only 32.04% went to the shoots and leaves. The findings showed that delayed application boosted the rate of fertiliser N consumption.

Increased fertiliser availability, use efficiency, and uptake, improved soil physico-chemical qualities, improved growth, and yield attributes, and greater HI (Harvest index) of economically valuable portions were the results of reducing RDF by 25% and replacing that 25% with VC [66]. By enhancing the availability and uptake of these nutrients, the application of 75% RDF and 25% organic manures raised the nutrient harvest indices in ginger. Higher crop harvest indices also indicate that the plant's economic portion contributed to a higher biomass production and, as a result, accumulated more nutrients than other portions. Under 75% RDF with organic manures, higher nutrient usage efficiency was seen in ginger [79].

Due to better nutrient availability, as well as the subsequent impact of integrated nutrient management on crop quality characteristics and higher yield over organic and control practices, higher nutrient removal and NPK uptake under integrated nutrient management practices over organic management were obvious. The direct input of N through inorganic fertilisers and its consistent availability from FYM and VC applied to the soil may be responsible for the increase in N uptake. The production of organic acids during the breakdown of organic manures may have helped to increase the solubility of both applied and native P, which may have contributed to the increase in P absorption under INM. The fact that the combination leads to a rise in root proliferation and, thus, higher nutrient uptake, may also serve as evidence for the higher nutrient uptake under INM. Additionally, INM's contribution to bettering soil aggregation would have resulted in a rise in root biomass and absorption rate. Using the full recommended amounts of NPK from organic sources resulted in noticeably low N, P, and K concentrations in ginger at harvest. This could be as a result of organic matter mineralizing slowly and less nutrients being available for crop growth and development. 35–50 kg P/ha are removed from a heavy ginger crop. Ca concentrations as low as 2 ppm are adequate to produce 90% of the maximum yield in the leaves of healthy ginger plants, which contain 1.1–1.3% Ca. Ref. [79] suggested using the fifth pair of leaves during the 90–120 DAP stage for foliar diagnosis of N, P, and K in order to determine the crop's nutritional needs.

8.3 Economics of INM in ginger

The primary cash crop for small farmers nationwide is ginger. Despite being one of the major industries in some parts of India, growing ginger has little knowledge regarding its economic feasibility and sustainability under integrated nutrient management. In comparison to organic nutrient management, integrated nutrient management systems produced higher values for a variety of economic factors. This demonstrates that organic nutrient management is the least profitable for farmers, which is clear given lower yields but higher input costs because more sources are needed to meet the nutrient need. Given its better yields, integrated nutrient management is undoubtedly advantageous. A combination of FYM with various doses of RDN in INM treatments produced greater BCRs (Benefit Cost Ratios), but lower net returns than VC. The results showed that replacing 25% of the fertiliser dose with VC would greatly benefit the farmers. Cheaper BCR was caused by the lower cost of FYM in comparison to VC, while greater NRR (Net Return Ratio) with VC integration could be attributed to significantly higher yields. At Pottangi, in the Indian state of Odisha, Azospirillum, FYM, and their combinations were studied for their effects. The application of Azospirillum at a rate of 10 kg/ha together with FYM at a rate of 10 t/ha resulted in the highest benefit-cost ratio of 2.4, according to ref. [68].

Based on major coefficient analysis, weighting, and ranking of various nutrition management methods, 100% RDN + RD of FYM received the top ranking because of its substantially greater SQI. Although 75% RDN + 25% N through VC on similar basis + pine mulch was given the second place because to increased biomass and rhizome yield, CQI, MRR, NUE, and harvest index. Thus, the optimum nutrient management module is created by saving 25% of fertiliser and subsequent input costs, which results in increased productivity. This is achieved by comparing the economic and efficiency superiority of 75% RDN + 25% N through VC to 100% RDN + RD of FYM [66].

9. Conclusion

The chapter has shown that the nutrient management systems has a significant impact on crop production, crop quality, soil health, and overall cultivation economics. In terms of yield, quality, utilisation efficiency and returns, the integrated method outperformed organic nutrient management significantly, increasing the overall output: input ratio and productivity. By increasing soil C, reduced bulk density, enhanced porosity, aggregation, and soil nutrient status, integrated nutrient management methods aid in improving and sustaining soil quality which has reported by several findings. As a result, this technology is superior and ideal for harvesting a bigger yield of ginger rhizomes of superior quality with profitable market returns.

Author details

N. Divyashree[1*], S. Poojashree[2], S. Venukumar[3] and Y.C. Vishwanath[1]

1 Department of Plantation, Spices, Medicinal and Aromatic Crops, College of Horticulture, University of Horticultural Sciences, Bagalkot, Karnataka, India

2 Department of Fruit Science, College of Horticulture, University of Horticultural Sciences, Bagalkot, Karnataka, India

3 Departmentof Entomology, College of Horticulture, University of Horticultural Sciences, Bagalkot, Karnataka, India

*Address all correspondence to: divyashreenbalaji@gmail.com

References

[1] Vasal PA. Ginger. In: Peter KV, editor. Handbook of Herbs and Spices. Vol. 1. New Delhi: Westville Publishing House; 2004

[2] Malu S, Obochi G, Tawo E, Nyong B. Antibacterial activity and medicinal properties of ginger (Zingiber officinale). Global Journal of Pure and Applied Sciences. 2009;**15**:3-4

[3] Jaidka M, Kaur R, Sepat S. Scientific Cultivation of Ginger (Zingiber Officinalis). New Delhi: Indian Agricultural Research Institute; 2018. pp. 110-112

[4] Srinivasan K. Ginger rhizomes (Zingiber officinale): A spice with multiple health beneficial potentials. Pharma Nutrition. 2017;**5**(1):18-28

[5] Masuda Y, Kikuzaki H, Hisamoto M, Nakatani NJB. Antioxidant properties of gingerol related compounds from ginger. Bio Factors. 2004;**21**(1-4):293-296

[6] Shukla Y, Singh M. Cancer preventive properties of ginger: A brief review. Food and Chemical Toxicology. 2007;**45**(5):683-690

[7] Ali BH, Blunden G, Tanira MO, Nemmar A. Some phytochemical, pharmacological and toxicological properties of ginger (Zingiber officinale roscoe): A review of recent research. Food and Chemical Toxicology. 2008;**46**(2):409-420

[8] Park YJ, Wen J, Bang S, Park SW, Song SY. Gingerol induces cell cycle arrest and cell death of mutant p53-expressing pancreatic cancer cells. Yonsei Medical Journal. 2006;**47**:688-697

[9] Ghasemzadeh A, Jaafar HZ, Rahmat A. Antioxidant activities, total phenolics and flavonoids content in two varieties of Malaysia young ginger (Zingiber officinale roscoe). Molecules. 2010;**15**(6):4324-4333

[10] Govindarajan VS. Ginger—Chemistry, technology, and quality evaluation: Part 1. Critical Reviews in Food Science and Nutrition. 1982;**17**:1-96

[11] Govindarajan VS. Ginger—Chemistry, technology, and quality evaluation: Part 2. Critical Reviews in Food Science and Nutrition. 1982;**17**:189-258

[12] Alqasoumi S, Yusufoglu H, Farraj A, Alam A. Effect of 6-shogaol and 6-gingerol on diclofenac sodium induced liver injury. International Journal of Pharmacology. 2011;**7**:868-873

[13] Sabina EP, Pragasam SJ, Kumar S, Rasool M. 6-gingerol, an active ingredient of ginger, protects acetaminophen-induced hepatotoxicity in mice. Journal of Chinese Integrative Medicine. 2011;**9**:1264-1269

[14] Chung WY, Jung YJ, Surh YJ, Lee SS, Park KK. Antioxidative and antitumor promoting effects of (6)-paradol and its homologs. Mutation Research. 2001;**496**:199-206

[15] Keum YS, Kim J, Lee KH, Park KK, Surh YJ, Lee JM, et al. Induction of apoptosis and caspase-3 activation by chemopreventive -paradol and structurally related compounds in KB cells. Cancer Letters. 2002;**177**:41-47

[16] Galal AM. Antimicrobial activity of 6-paradol and related compounds. Pharmaceutical Biology. 1996;**34**:64-69

[17] Ling H, Yang H, Tan SH, Chui WK, Chew EH. 6-Shogaol, an active

constituent of ginger, inhibits breast cancer cell invasion by reducing matrix metalloproteinase-9 expression via blockade of nuclear factor-κB activation. British Journal of Pharmacology. 2010;**161**:1763-1777

[18] Choudhury D, Das A, Bhattacharya A, Chakrabarti G. Aqueous extract of ginger shows antiproliferative activity through disruption of microtubule network of cancer cells. Food Chemical Toxicology. 2010;**48**:2872-2880

[19] Weng CJ, Wu CF, Huang HW, Ho CT, Yen GC. Anti-invasion effects of 6-shogaol and 6-gingerol, two active components in ginger, on human hepatocarcinoma cells. Molecular Nutrition & Food Research. 2010;**54**:1618-1627

[20] Shin SG, Kim JY, Chung HY, Jeong JC. Zingerone as an antioxidant against peroxynitrite. Journal of Agricultural and Food Chemistry. 2005;**53**:7617-7622

[21] Aeschbach R, Löliger J, Scott BC, Murcia A, Butler J, Halliwell B, et al. Antioxidant actions of thymol, carvacrol, 6-gingerol, zingerone and hydroxytyrosol. Food Chemical Toxicology. 1994;**32**:31-36

[22] Chung SW, Kim MK, Chung JH, Kim DH, Choi JS, Anton S, et al. Peroxisome proliferator-activated receptor activation by a short-term feeding of zingerone in aged rats. Journal of Medicinal Food. 2009;**12**:345-350

[23] Kim MK, Chung SW, Kim DH, Kim JM, Lee EK, Kim JY, et al. Modulation of age-related NF-kappaB activation by dietary zingerone via MAPK pathway. Experimental Gerontology. 2010;**45**:419-426

[24] Manjunatha JR, Bettadaiah BK, Negi PS, Srinivas P. Synthesis of quinoline derivatives of tetrahydrocurcumin and zingerone and evaluation of their antioxidant and antibacterial attributes. Food Chemistry. 2013;**136**:650-658

[25] Kirana C, McIntosh GH, Record IR, Jones GP. Antitumor activity of extract of Zingiber aromaticum and its bioactive sesquiterpenoid zerumbone. Nutrition and Cancer. 2003;**45**:218-225

[26] Abdul AB, Abdelwahab SI, Al-Zubairi AS, Elhassan MM, Murali SM. Anticancer and antimicrobial activities of zerumbone from the rhizomes of Zingiber zerumbut. International Journal of Pharmacology. 2008;**4**:301-304

[27] Lee HY, Park SH, Lee M, Kim HJ, Ryu SY, Kim ND, et al. 1-Dehydro-[10]-gingerdione from ginger inhibits IKKβ activity for NF-κB activation and suppresses NF-κB-regulated expression of inflammatory genes. British Journal of Pharmacology. 2012;**167**:128-140

[28] Dinesh R, Srinivasan V, Hamza S, Singh HP, Parthasarathy VA, Kandiannan K, et al. Nutrition. In: Zingiberaceae Crops—Present and Future. New Delhi: Westville Publishing House. pp. 255-287

[29] Rahman S, Salehin F, Iqbal A. In vitro antioxidant and anticancer activity of young Zingiber officinale against human breast carcinoma cell lines. BMC Complementary Medicine. 2011;**11**:76

[30] Gowariker V, Krishnamurthy VN, Gowariker S, Dhanorkar M, Paranjape K, Borlaug N. The Fertilzer Encyclopedia. John Wiley and Sons, Technology & Engineering; 2009. p. 880

[31] Liu Y, Whelan RJ, Pattnaik BR, Ludwig K, Subudhi E, Rowland H, et al. Terpenoids from Zingiber officinale (ginger) induce apoptosis in endometrial

cancer cells through the activation of p53. PLoS One. 2012;7:53-178

[32] Dinesh R, Srinivasan V, Hamza S, Manjusha A. Short-term incorporation of organic manures and biofertilizers influences biochemical and microbial characteristics of soils under an annual crop [turmeric (Curcuma longa L.)]. Bioresource Technolology. 2010;**101**:4697-4702

[33] Azeze S, Naruka IS, Singh PP, Kushwah SS. Nutrient management and its effect on growth, yield and quality of ginger cultivars. Indian Journal of Horticulture. 2013;**70**:65-70

[34] Li L, Chen F, Yao D, Wang J, Ding N, Liu X. Balanced fertilization for ginger production—Why potassium is important. Better Crops. 2010;**94**:25-27

[35] Sharath Pal MV, Hegde NK, Hanamashetti SI, Kulkarni MS. Effect of organic manures on the performance of ginger under northern dry zone of Karnataka. Journal of Spices and Aromatic Crops. 2014;**23**:121-124

[36] Singh AK, Gautam US, Singh J. Impact of integrated nutrient management on ginger production. Bangladesh Journal of Botany. 2015;**44**:341-344

[37] Singh SP. Nutrient supplementation through organic manures for growth and yield of ginger (Zingiber officinale rose.). Journal of Eco-Friendly Agriculture. 2015;**10**:28-31

[38] Loganath M, Das NK. Determination of optimum spacing, fertilization and method of planting for ginger (Zingiber officinale Rosc.). Indian Journal of Agronomy. 1964;**9**:281-285

[39] Thomas KM. Influence of N and P2O5 on the yield of ginger. Madras Agricultural Journal. 1965;**52**:512-515

[40] Aiyadurai SG. A Review of Research on Spices and Cashewnut in India. Ernakulam: Regional Office (Spices and Cashewnut), ICAR; 1966. p. 228

[41] Nair GS. Note on the use of detached sprouts as a planting material in ginger. Agricultural Research Journal of Kerala. 1977;**15**(1):100-101

[42] Randhawa KS, Nandpuri KS. Yield and quality of ginger (Zingiber officinale Rosc.) varieties grown in Assam. The Punjab Horticultural Journal. 1970;**10**:111-112

[43] Nair KP. Ginger nutrition. In: Turmeric (Curcuma longa L.) and Ginger (Zingiber officinale Rosc.)-World's Invaluable Medicinal Spices. Cham: Springer; 2019. pp. 433-440

[44] Nybe EV, Raj NM. Ginger production in India and other South Asian countries. In: Ginger. CRC Press; 2016. pp. 231-260

[45] Haque MM, Rahman AKMM, Ahmed M, Masud MM, MMR S. Effect of nitrogen and potassium on the yield and quality of ginger in hill slope. Journal of Soil and Nature. 2007;**1**:36-39

[46] Xizhen A, Jinfeng S, Xia X. Ginger production in Southeast Asia. In: Ginger. Boca Raton, USA: CRC Press; 2016. pp. 261-298

[47] Tahat MM, Sijam K. Mycorrhizal fungi and abiotic environmental conditions relationship. Research Journal of Environmental Sciences. 2012;**6**:125-133

[48] Maherali K. Influence of phylogeny on fungal community assembly and ecosystem functioning. Science. 2007;**316**:1746

[49] Sudha B, John J, Meera AV, Sajeena A. Growth, nutrient uptake and yield of ginger as impacted by potting media,

foliar nutrition and microbial inoculants. Journal of Spices & Aromatic Crops. 2020;**29**:2-5

[50] Sharma S, Dohroo NP, Korla BN. Effect of VAM inoculation and other field practices on growth parameters of ginger. Journal of Hill Research. 1997;**10**:74-76

[51] Singh SP, Dwivedi DK. Impact of zinc, boron and iron elements on yield and economics of ginger. International Journal of Agricultural Sciences. 2007;**3**:136-138

[52] Doolette CL, Read TL, Li C, Scheckel KG, Donner E, Koppittke PM. Foliar application of zinc sulphate and zinc EDTA to wheat leaves: Differences in mobility, distribution and speciation. Journal of Experimental Botany. 2018;**69**:4469-4481

[53] Prasad R. Effect of Foliar Spray of Urea and NAA on the Growth and Yield of Ginger (Zingiber officinale Roscoe). 2021

[54] Shadap A, Pariari A, Lyngdoh YA. Effect of foliar sprays of micronutrients on the performance of ginger (Zingiber officinale Rosc.). Annals of Biology. 2018;**34**(2):212-214

[55] Nayak RJ, Ravi CS, Ganapathi M, Shivaprasad M, Bhoomika HR. Influence of foliar application of benzyl adenine and nutrients on growth and yield of transplanted ginger (Zingiber officinale rosc.) under hill zone of Karnataka. International Journal of Current Microbiology and Applied Sciences. 2020;**9**(9):1793-1798

[56] Prajapati RK, Nongrum K, Singh L. Growth and productivity of ginger (Zingiber officinale Rosc.) under kapok (Ceiba pentandra L. Gaertn) based Agri–silviculture system. Indian Journal of Agroforestry. 2007;**9**:12-19

[57] Shaikh AA, Desai MM, Shinde SB, Tambe AD. Yield and nutrient uptake of ginger (Zingiber officinale Rosc.) as affected by organic manures and fertilizers. International journal of. Agricultural Sciences. 2010;**6**(1): 28-30

[58] Sreekala GS, Jayachandran BK. Effect of organic manures and microbial inoculants on nutrient uptake, yield and nutrient status of soil in ginger intercropped coconut garden. Journal of Plantation Crops. 2006;**34**:25-31

[59] Cho GH, Yoo CH, Choi JW, Park KH, Hari SS, Kim SJ. Research report rural development administration, plant environment mycology and farm products utilisation. Korea Republic. 1997;**29**:30-42

[60] Chengat T. Influence of organic manures and Azospirillum on growth, yield and quality of ginger (Zingiber officinale). M.Sc. (Hort.) Thesis, Kerala Agricultural University, Kerala, India. 2014

[61] Pariari A, Karthik CS, Bhattacharya S. Effect of organic manures on growth, yield and quality of mango ginger (Curcuma amada Roxb.) in the Gangetic Alluvial Plains of West Bengal. International Journal of Current Microbiology and Applied Sciences. 2019;**8**(11):1030-1034

[62] Egbuchua CN, Enujeke EC. Growth and yield responses of ginger (Zingiber officinale) to three sources of organic manures in a typical rainforest zone. Nigeria. Journal of Horticulture and Forestry. 2013;**5**(7):109-114

[63] Srinivasan V, Shiva KN, Kumar A. Ginger. In: Parthasarathy VA, Kandiannan K, Srinivasan V, editors. Organic Spices. New Delhi: New India Publishing Agency; 2008. pp. 335-386

[64] Yadav RK, Yadav DS, Rai N, Sanwal SK, Sarma P. Commercial prospects of ginger cultivation in north-eastern region. ENVIS Bulletin: Himalayan Ecology. 2004;**12**(2):1-5

[65] Lester GE, Saftner RA. Organically versus conventionally grown produce: Common production inputs, nutritional quality, and nitrogen delivery between the two systems. Journal of Agricultural and Food Chemistry. 2011;**59**(19):10401-10406

[66] Worthington V. Nutritional quality of organic versus conventional fruits, vegetables, and grains. The Journal of Alternative and Complementary Medicine. 2001;7(2):161-173

[67] Kakar R, Sharma JC, Mogta A, Guleria A, Thakur J. Assessment of various nutrient management technologies for quality, fertilizer use efficiency, and economics of ginger production under subtropical to subtemperate conditions. Communications in Soil Science and Plant Analysis. 2020;**51**(22):2805-2820

[68] Dash DK, Mishra NC, Sahoo BK. Influence of nitrogen, Azospirillum sp. and farm yard manure on the yield, rhizome rot and quality of ginger (Zingiber officinale Rosc.). Journal of Spices and Aromatic Crops. 2008;**10**:177-179. Special issue on Proceedings of the National Symposium on Spices and Aromatic Crops: Threats and solutions to spices and aromatic crops industry

[69] Mei LZ, Lan LL, Yu SC, Guang CX, Chao Z, Lan YH. Effects of humic acid fertilizer on urease activity in ginger growing soil and nitrogen absorption of ginger. China Vegetables. 2009;**4**:44-47

[70] Mohanty DC, Naik BS, Panda BS. Ginger research in Orissa with reference to its varietal and cultural improvement. Indian Cocoa Arecanut Spices Journal. 1990;**14**:61-65

[71] Pawar HK, Patil BR. Maharashtra Agricultural. University Journal. 1987;**12**(3):350-354

[72] Dinesh R, Srinivasan V, Hamza S, Manjusha A, Sanjay KP. Short-term effects of nutrient management regimes on biochemical and microbial properties in soils under rainfed ginger (Zingiber officinale Rosc.). Geoderma. 2012;**173-174**:192-198

[73] Srinivasan V, Thankamani CK, Dinesh R, Kandiannan K, Hamza S, Leela NK, et al. Variations in soil properties, rhizome yield and quality as influenced by different nutrient management schedules in rainfed ginger. Agricultural Research. 2019;**8**:218-230

[74] Bhattacharya SS, Kim K-H, Das S, Uchimiya M, Jeon BH, Kwon E, et al. A review on the role of organic inputs in maintaining the soil carbon pool of the terrestrial ecosystem. Journal of Environmental Management. 2016;**167**:214-227

[75] Ayuba SA, John C, Obasi MO. Effects of organic manure on soil chemical properties and yield of ginger research note. Nigerian Journal of Soil Science. 2005;**15**:136-138

[76] Nagarajan M, Pillai NG. Note on nutrient removal by ginger and turmeric rhizomes. The Madras Agricultural Journal. 1979;**66**:56-59

[77] Haag HP, Saito S, Dechen AR, Carmello QAC. Anais da Escola Superior de agriculture. Luiz de Queiroz. 1990;**47**:435-457

[78] Xu K, Guo YY, Wang XF. Transportation and distribution of carbon and nitrogen nutrition in ginger. Acta Horticulture. 2004;**629**:347-353

[79] Johnson PT. Foliar Diagnosis, Yield and Quality of Ginger in Relation to N, P and K, M.Sc. (Agric.) Thesis. Kerala, India: Kerala Agricultural University; 1978

Chapter 9

Ginger Based Agro-Forestry Systems for Livelihood to Rainfed Areas

Rakesh Chandra Nainwal and Shri Krishna Tewari

Abstract

In rural areas particularly belonging to tropical rainfed zone, agro-forestry is a very common strategy adopted as a common popular tool for saving environmental degradation, in which multipurpose trees (MPTs) are planted with common agriculture crops. These MPTs play also a very vital role for rehabilitating degraded lands and enhancing the total productivity of the land with proper combination of these MPTs with different crops. Such kinds of agroforestry systems provide livelihood security to the farmers of rainfed areas. In India, ginger is planted as intercrop with different tree species and being a shed-loving plant, and its yield was increased as compared with monoculture system.

Keywords: agroforestry, ginger, livelihood, rainfed

1. Introduction

Rainfed areas are mainly dependent on uncertain rainfall which generally deficient to water requirements [1] and vagaries of weather [2]. In India a large part nearly 57 per cent of the agricultural land spread across the country, comes under rainfed areas and hence considered very important for agricultural productivity and livelihood for millions of rural households. Generally rainfed areas are lacking or short of some important natural resources mainly like water availability and critical environment, which also leads to land degradation due to water and wind erosion, also lower water use efficiency (WUE) resulted in low economic yield of field crops [3–5]. However, if these areas managed properly they also can share in good amount in food grain production. These high potential areas provide opportunities for improvement in agriculture production.

The primary reason for land degradation in rainfed areas is poor land use and disordered agriculture production leads to shortage of the natural resources. There are some following natural resources and their relationship between soil and water degradation [6] need to be focused [6], are:

1. Soil organic matter loss and physical degradation: Soil organic matter plays a very crucial role to managing water cycles in any of ecosystem. Decayed in level

of organic matter have significant effect on soil physical properties like negative impacts on infiltration, porosity and water holding capacity of the soil.

2. Chemical degradation and nutrient depletion of soil: In rainfed areas, due to lack of proper water sufficiency and distribution causes the imbalance of essential nutrients in agricultural soil and making the soil deficient particularly in terms of macro nutrients like N, P and K [7]. Sometime, imbalance of nutrients also leads to salt stress problems like salinity and alkalinity due to accumulation of Na salts on soil surface.

3. Soil erosion and sedimentation: Surface soil loss through erosion due to high thunder storm and poor soil aggregate stability and vegetative cover, is an extreme problem of rainfed areas [8]. These erosion activities lead to nutrient loss from upper soil layer resulting lower soil productivity in these areas. It also results in high sediment yields in downstream areas, and negative environmental impact [9].

4. Water scarcity: Sometimes due to high and increase in dry spell creates to drought like conditions in rainfed areas [10] resulting in decreased water availability and negative impact on crop growth and yield. Additionally, the combined effect of higher temperature and low water availability leads to higher crop evaporation demand. Consequentially higher temperature results in lower crop growth, yield and productivity [11].

2. Socio economic status of rainfed area

At present around 55 percent area of net sown area of the country comes under rainfed, which is considered as home for around 65–70% of total livestock and 40–50% of total human population. The poor socio economic condition of these areas comprising low irrigation facility as compared to irrigated regions, lower productivity and lesser employment opportunities leads to migration of the people for their livelihood (**Table 1**).

In rainfed areas for sustainable development, there is a considerable scope for land use diversification and crop intensification in areas having high NRI (Natural Resource Index) and low or medium ILI (Integrated Livelihoods Index). Crop production in uncertain rainfall areas is risky. Low and unstable yields are common and so is the income of dryland farmers. For imparting stability and providing sustainability to the farming systems, a tree-crop (Agroforestry) integration holds promise. Agroforestry systems can meet the multiple requirements of food, fodder, fuel, fertilizer, etc., besides improved pasture management.

Agroforestry is a multifarious approach for sustainable utilization of land and other natural resources by incorporating the trees in to diverse agricultural farming systems on same land and time to achieve economic, environmental, ecological, and cultural benefits [12]. An agroforestry system has three main objectives [13]:

- Protecting and stabilizing impact on the ecosystems;

- Producing a high level of output of economic goods and services;

- Improving income and employment to rural population

Parameter	Rainfed region	Irrigated region
Poverty ratio (%)	37	33
Proportion of agricultural labor (%)	30	28
Land productivity (INR/ha)	5716	8017
Proportion of irrigate area (%)	15	48
Per capita consumption (kg/year) of:		
Cereals	240	459
Pulses	20	12
Total food grains	260	471
Cooperative credit (INR/ha)	816	1038
Infrastructure development index	0.30	0.40
Social development index	0.43	0.44

Source 1. GoI (2006)-NSSO 2006, p. 18.

Table 1.
Comparison of rainfed vis-à-vis irrigated regions.

And to fulfill these objectives, there is a need to elucidate a proper combination of crop and tree, which must be complimentary or supplementary among the components in the system.

Ginger is the most suitable spice/aromatic shade loving crop for intercropping in agro-forestry systems in north regions from lowlands (500 mt.) to medium elevation (500–1000 mt.) [14]. Ginger, (*Zingiber officinale*), belongs to the family Zingiberaceae, is an herbaceous perennial commercial plant, used as a spice, flavoring, food and medicine. Besides consuming fresh ginger as a vegetable, several value added products are prepared from fresh ginger such as ginger paste, ginger candy and essential oils and Oleoresins. Ginger paste offers convenience to consumer's along with easy storage, long shelf life and authentic taste, to suit the requirements of consumers. There is a huge demand of such type of packaged ginger products in commercial shops like hotels and restaurants specially. In the same way, the demand of ginger oil and oleoresins is also raising in food processing, pharmaceutical and nutraceutical sector.

Ginger is an important commercial horticultural crop, and is cultivated in more than 35 countries around the world. In 2019, the global production of ginger was 4,081,374 tons. India is among the top leading producer of ginger in the global economy, having 45 percent share in area and 35.2 percent share in production and Nigeria was second, accounting for 16.94% [15]. The area under ginger in India was 53,900 hectares in 1990–91, showed a gradual increase over the years, and was 172,040 hectares in 2019–20 [16]. Recently the ginger paste and oil, has raised its demand in Ayurvedic system of medicine, due to its carminative and stimulant properties. A number of attempts have been made from last two-three decades to sustain its productivity by screening of new and high yielding varieties, optimization of nutrient doses, improved agronomic practices and protection measure [17]. The shade-loving nature of ginger [17, 18] have encouraged the researchers to explore its potential as an intercrop and many intercropping models have also been developed with ginger under different tree canopies viz., teak [18], coconut [19], areca nut [20], paulownia [21] and ailanthus [22].

3. Trees for plantation in agroforestry

In dryland or rainfed regions trees are planted on field bunds or boundaries as live fences or serving as wind break, to utilize the space. For this purpose, mostly those trees are planted with straight growth habit to avoid the interference and disturbance by their shading to the associated crop [23, 24]. The common tree species grown as boundary plantations in dry land systems are *Tectona grandis*, *Leucaena leucocephala* (pollarded for fodder), *Borassus flabellifer*, *Cocos nucifera*, *Acacia nilotica* var. cupressiformis, *Dalbergia sissoo* and *Prosopis juliflora* [23, 24]. In the last few years, systematic experiments are being conducted at several locations of India covering aspects on agroforestry, as a result of which several recommendations have emerged [25] for different agro-climatic regions of the country for rainfed conditions (**Table 2**).

In the era of climate change, accommodation of ginger as an intercrop in diverse agroforestry models can be a vital production system for environmental amelioration, realizing higher monetary returns and for sustaining the fertility of the soil. The various study inferred that ginger yield, soil physico-chemical and nutrient contents were higher [27] when grown in association with any plant spp. in agroforestry system (**Table 3**) as compared to sole cropping (**Table 4**). For effective, beneficial and

Traditional farm forestry	*Acacia nilotica, Albizia odoratissima, Annona squamosal, Azadirachta indica, Butea monosperma, Gmelina arborea, Mangifera indica, Tectona grandis, Syzigium cuminii, Terminalia bellirica, Terminalia arjuna, Ziziphus mauritiana*
Farm boundary plantations	*A. catehu, A. indica, Albizia lebbeck, A. nilotica, A. procera, Bambusa arundinacea, B. vulgaris Dalbergia sissoo, Dendrocalamus strictus, Eucalyptus* spp.*, Gliricidia maculata, G. arborea, Leucaena Leucocephala, Pongamia pinnata, T. grandis*
Block plantations/Farm wood lots	*A.mangium, Casuarina Equisetifolia, D. sissoo, D. strictus, Eucalyptus spp., L. leucocephala, T. grandis,*
Natural silvopasture	*A. lebbeck, A. nilotica, Annona Squamosal, Erythrina Indica, Emblica Officinalis, Hardwickia binate,, M. indica, Z. mauritiana*
Live hedges	*A. Senegal, P. juliflora, Bamboo spp., CaesalpiniaSepiaria, Dodonaea Viscosa, Ipomoea carnia, Lawsonia Inermis, Lantana camara, Vitexnegundo, Z. oenoplia*

Source: [26]

Table 2.
List of some woody species integrated in Agroforestry system of dry land.

Items	**Agroforestry models**		
	Acacia – Pineapple (TK)	**Acacia – Ginger (TK)**	**Acacia – Turmeric (TK)**
Total gross income	442,900	665,100	413,507
Total production cost	201,000	239,244	214,726
Net income	241,900	425,856	198,781
BCR	2.20	2.78	1.93

**TK = taka (Bangladeshi currency).*

Table 3.
Cost of production, total income and net income of diverse agroforestry models in hectare per year.

Treatment	Garlic production	Net return (Tk/ha)	BCR
$S_1 + G_1$ + Garlic	30,900	259,473	4.15
$S_1 + L_1$ + Garlic	44,625	234,905	3.99
$S_2 + G_2$ + Garlic	40,230	226,654	4.11
$S_2 + L_2$ + Garlic	66,150	219,498	4.12
$S_3 + G_3$+ Garlic	50,317	176,979	3.90
$S_3 + L_3$+ Garlic	78,635	181,945	4.04
Open	182,400	120,860	3.14

S_1 = Dalbergia Sissoo (spacing 4 m × 4 m); S_1 = D. sissoo (spacing 5 m × 5 m); S_1 = D. sissoo (spacing 5 m × 5 m); G_1 = guava (spacing 2 m × 2 m); G_2 = guava (spacing 2.5 m × 2.5 m); G_3 = guava (spacing 3 m × 3 m); L_1 = lemon (spacing 2 m × 2 m); L_1 = lemon (spacing 2.5 m × 2.5 m); L_1 = lemon (spacing 3 m × 3 m); Open = sole cropping of ginger; BCR = benefit to cost ratio.
(Source: [28])

Table 4.
Economics of ginger production under Dalbergia Sissoo based multistrata cropping system.

Treatments	Total cost of production (₹/ha)	Net income (₹/ha)	BCR
Sapota + Jatropha + Ginger (var. Navsari local)	159343.88	240598.65	2.52
Sapota + Jatropha + Ginger (var. Udaipur local)	159343.88	279293.65	2.76
Sapota + Ginger (var. Navsari local)	159343.88	95755.65	1.60
Sapota + Ginger (var. Udaipur local)	159343.88	136631.65	1.87
Jatropha + Ginger (var. Navsari local)	159343.88	171396.15	2.08
Jatropha + Ginger (var. Udaipur local)	159343.88	222063.65	2.40
Ginger (var. Navsari local)	152323.88	59698.63	1.40
Ginger (var. Udaipur local)	152323.88	105031.13	1.70

(Source: [29])

Table 5.
Comparative economics of Ginger based agroforestry models.

compatible cultivation of ginger as intercrop in any agroforestry model, multistrata system can also be a highly remunerative approach (**Table 4**) to overcome a number of problems like to poverty by improving the socio-economic status and reducing the side effect of global warming. Research revealed ginger based cropping models result in higher net income and benefit cost ratio [29] with lower cost of production (**Table 5**).

4. Factor affecting ginger production in agri-silviculture system

Temperature: Ginger prefers the warm and humid climate with the temperature 22–28°C,at seedling stage and 25°C during rhizome enlarging stage [30] however, also adaptable to low temperature. Rising in temperature increase the germination but weakens the sprouting of the rhizome.

Light: Ginger comes under category of shade loving plants and grows well I low to medium light intensity. During germination stage it needs darkness while medium light at seedling stage and high light during growing stage [31].

Water: Ginger plant cannot withstand in water stress condition as the soil water availability directly affect the rhizome growth [32]. For optimum growth of the plant well drained soil with proper irrigation facility is suitable.

Seed size: The growth of seedling largely depends upon the seed size as higher the seed size increases the sprouting and grow vigorously. However too big seed size should always be avoided.

Spacing: Ginger is one of the best suited crops for intercropping due to its shade loving nature. It needs around 30–40% illumination through radiation for optimum growth of the plant and rhizome as well [17]. For intercropping 5 m x 3-4 m spacing is found best spacing for rhizome production in agroforestry system.

Fertilizer: During the whole life cycle ginger needs changing NPK proportion at its different stages [33] as in early stage it needs more K followed by N and P while later in stage K uptake is lesser and N and P uptake is more for its luxurious growth and higher rhizome yield. However, during its growth period ginger need NPK in proportion of 11:1:16.1.

Although ginger often gains attention for its potential to generate high profits for farmers, limited attention has been given to expanding production aspects for the betterment of smallholder farmers engaged in production and marketing activities. In the climate change era, ginger based agroforestry models can be a vital production system for environmental amelioration, realizing higher monetary returns and for sustaining the fertility of the soil. The study inferred that ginger yield, soil physico-chemical and nutrient contents were higher when grown as intercrop sole cropping. Overall it can be concluded that there is tremendous potential for ginger cultivation to strengthen the agricultural sector by adopting suitable strategies. But this needs to happen scientifically at every stage from production to post harvest and processing stage. There is scope to widen markets in both domestic and international markets for this crop as well as its value additions.

Author details

Rakesh Chandra Nainwal* and Shri Krishna Tewari
Plant Conservation and Agro-Technology, CSIR-National Botanical Research Institute, Lucknow, India

*Address all correspondence to: nainwal.rakesh@gmail.com

References

[1] Srinivas K, Vittal KPR, Sharma KL. Resource characterization of dryland: Soils. In: Singh HP, Ramakrishna YS, Sharma KL, Venkateswarlu B, editors. Fifty Years of Dryland Agriculture Research in India. Hyderabad, India: Central Research Institute for Dryland Agriculture; 1999. pp. 41-55

[2] Ramakrishna YS, Vijaya KP, Ramana RBV. Crop-weather relationship studies in dryland agriculture. In: Singh HP, Ramakrishna YS, Sharma KL, Venkateswarlu B, editors. Fifty Years of Dryland Agriculture Research in India. India: Central Research Institute for Dryland Agriculture; 1999. pp. 211-226

[3] Rockström J, Hatibu N, Oweis T, Wani SP. Managing water in rain-fed agriculture. In: Molden D, editor. Water for Food, Water for Life: A Comprehensive Assessment of Water Management in Agriculture. London, UK: Earthscan; 2007. pp. 315-348

[4] Wani SP, Pathak P, Sreedevi TK, Singh HP, Singh P. Efficient management of rainwater for increased crop productivity and groundwater recharge in Asia. In: Kijne JW, Barker R, Molden D, editors. Water Productivity in Agriculture: Limits and Opportunities for Improvement. Wallingford, UK: CAB International; 2003b. pp. 199-215

[5] Wani SP, Singh HP, Sreedevi TK, Pathak P, Rego TJ, Shiferaw B, et al. Farmer participatory integrated watershed management: Adarsha watershed, Kothapally, India. An innovative and up-scalable approach. A case study. In: Harwood RR, Kassam AH, editors. Research Towards Integrated Natural Resources Management: Examples of Research Problems, Approaches and Partnerships in Action in the CGIAR. Washington, DC: Interim Science Council, Consultative Group on International Agricultural Research; 2003c. pp. 123-147

[6] Bossio D, Critchley W, Geheb K, van Lynden G, Mati B. Conserving land – Protecting water. In: Molden D, editor. Water for Food, Water for Life: A Comprehensive Assessment of Water Management in Agriculture. London, UK: Earthscan; 2007. pp. 551-584

[7] Irshad MM, Inoue M, Ashraf HK, Faridullah M, Delower A, Tsunekawa. Land desertification-an emerging threat to environment and food security of Pakistan. Journal of Applied Sciences. 2007;**7**(8):1199-1205

[8] Huang GB, Zhang RZ, Li GD, Li LL, Chan KY, Heenan DP, et al. Productivity and sustainability of a spring wheat-field pea rotation in a semi-arid environment under conventional and conservation tillage systems. Field Crops Research. 2008;**107**:43-55

[9] Huang GB. No-till and reduced tillage test and application in Gansu province. In: Paper Presented at the Conservation Tillage Workshop Organized by the Center for Conservation Tillage Research (CCTR) with the Support of ACIAR. Lanzhou, China; 2003

[10] Sillmann J, Roeckner E. Indices for extreme events in projections of anthropogenic climate change. Climatic Change. 2008;**86**:83-104

[11] Food and Agriculture Organization of the United Nations (FAO). Climate Change, Water and Food Security. Rome, Italy: FAO; 2011

[12] Thevathasan NV, Gordon AM, Simpson JA, Reynolds PE, Price GW,

Zhang P. Biophysical and ecological interactions in a temperate tree-based intercropping system. Journal of Crop Improvement. 2004;**12**:339-363

[13] Chauhan SK, Dillon WS, Singh N, Sharma R. Physiological behaviour and yield evaluation of agronomic crops under Agri-horti-silviculture system. International Journal of Plant Research. 2013;**3**:1-8

[14] Nair PKR. An Introduction to Agro Forestry. India: Springers; 1993. p. 200

[15] FAOSTAT. Production quantities of Ginger by country 2019. 2020b. Retrieved from: http://www.fao.org/faostat/en/#data/QC/visualize

[16] Shroff S. Assessment of Ratio of Different Products/Forms of Spices Being Marketed: Study Based on Ginger and Turmeric. AERC Report, 2020. Pune, India: Agro Economic Research Centre (AERC). Gokhale Institute of Politics and Economics; 2020

[17] Jaswal SC, Mishra VK, Verma KS. Intercropping ginger and turmeric with poplar (*Populus deltoides* 'G-3' marsh). Agroforestry Systems. 1993;**22**:111-117

[18] Lahiri AK. Intercropping trials with turmeric in North Bengal. Indian Forester. 1972;**98**:109-115

[19] Satheesan KV, Ramadasan A. Growth and productivity of turmeric grown as a pure and as an intercrop in coconut garden. In: The Proceeding of the National Seminar on Ginger and Turmeric. Calicut, India: CPCRI; 1980. pp. 69-75

[20] Singh RK, Bhalerao MM, Reddy MJM. Costs and returns of intercrops in Arecanut. Indian Cocoa, Areca nut and Spices Journal. 1986;**9**:99

[21] Newman SM, Bennett K, Wu Y. Performance of maize, beans and ginger as intercrops in Paulownia plantations in China. Agroforestry Systems. 1998;**39**:23-30

[22] Kumar BM, Thomas J, Fisher RF. *Ailanthus triphysa* at different density and fertilizer levels in Kerala, India: Tree growth, light transmittance and understory ginger yield. Agroforestry Systems. 2001;**52**:133-144

[23] Korwar GR. Alternate land use systems: Trees and bushes. In: Singh HP, Ramakrishna YS, Sharma KL, Venkateswarlu B, editors. Fifty Years of Dryland Agriculture Research in India. India: Central Research Institute for Dryland Agriculture; 1999. pp. 507-512

[24] Pathak PS, Pateria HM, Solanki KR. Agroforestry Systems in India - a Diagnosis and Design Approach. Jhansi, India: ICAR; 2000

[25] Pathak PS, Solanki KR. Agroforestry Technologies for Different Agro-climatic Regions of India. Jhansi, India: ICAR; 2002. p. 41

[26] Ilorkar VM, Suroshe SB, Jiotode DJ. Agroforestry interventions across different agro-climatic zones in Maharashtra, India. Indian Journal of Forestry. 2011;**34**(1):105-109

[27] Chauhan HS, Kamla S, Patra DD, Singh K. Studies on litter production, nutrient recycling and yield potential under (5-6 years old) poplar (*P. deltoides*) and Ecualyptus (E, hybrid) interplanted with aromatic crops in Tarai region of Uttar Pradesh. Journal of Medicinal Aromatic Plant Science. 1997;**19**:1034-1038

[28] Bari MS, Rahim MA. Production potential of ginger under different spacing of *Dalbergiasissoo*. Journal

of Agroculture Environment. 2010;**4**(1):143-146

[29] Pandey SBS, Jadeja DB, Manohar NS, Tandel MB. Economic comparison of intercropping of ginger and turmeric under sapota-jatropha based agro-forestry systems in South Gujrat. International Journal of Science, Environment and Technology. 2016;**5**(5):3635-3642

[30] Xizhen A, Zhenxian Z, Shaohui W. Effect of temperature on photosynthetic characteristics of ginger leaves. China Vegetables. 1998;**3**:1-3

[31] Dewan Z, Kun X, Liping C. Study on photosynthetic characteristics of ginger. Acta Hort Sinica. 1991;**18**:55-60

[32] Kun X. The influences of mulching with straw on the field microclimate and ginger growth. China Vegetables. 1999;**5**(5-8):14

[33] Xu K, Limei K, Zhao D. Uptake and Distribution of NPK by Ginger Plants. Shandong, China: Shandong Agricultural Sciences; 1992

Chapter 10

Cumin (*Cuminium cyminium* L.): A Seed Spice Crop with Adopted Production Technology in Cumin Cultivated Regions

Navjot Singh Brar, Prakash Mahala, Kartik Sharma, Parmdeep Singh Dhanda, Alpa Yadav, Meenakshi Sharma and Prashant Kaushik

Abstract

Cumin is a seed spice which finds its place in variety of global cuisines, especially in Indian context. India leads in the world in production of cumin with 70% of world's production and consumes 90% of this produce. It is a high potential crop with great demand around the world due to changing food consumption behavior, and increasing demand for value-added products such as oil and powder. Cumin has a distinct flavor and aroma owing to presence of essential oils. Cumin has different biological and biomedical properties and finds use in various ayurvedic preparations in different forms. Cumin has been found in three types of colours: amber, white, and black. Among this amber is widely accepted and black also have unique flavor. Cumin is a crop of tropical and subtropical regions and suitable for cultivation on wide variety of soils. Cumin production can be easily done with very few hindrances such as frost injury, wilt and powdery mildew. There is a lot of scope and prospectus regarding its cultivation which can be exploited in other cumin suitable regions of the world through various agronomical innervations, crop improvement programs and biotechnological tools.

Keywords: cumin, seed, spice, climate, soil

1. Introduction

The cumin (*Cuminium cyminium* L.) commonly known as *Jeera* is an important seed spice crop belonging to family Apiaceae of the order Apiales. It holds more than 22% share of area under spice crops, and is the most widely grown seed spice in India with an area share of more than 48% among seed spices. In India cultivation of this spice is very popular in states of Rajasthan and Gujarat, along with some parts of Madhya Pradesh and Uttar Pradesh during *Rabi* season. During 2019–20, cumin was cultivated on 8.42 lakh hectares in India with total production amounting to 5.47 lakh tonnes. Rajasthan and Gujarat are the leading producers and contribute to nearly

99% of the total production of India. India as a leading producer has a world market share of 70%, followed by Syria (13%), Turkey (5%), and UAE (3%) [1]. These four countries produce about 91% of cumin of the world, while the remaining production comes from other tropical or sub-tropical Asian and African countries.

Cumin as a condiment is vital in composition of mixed spices, curry powders, and it also imparts good flavour to soups, sausages, pickles etc. Some of the bakery products such as breads and cakes are also seasoned with these seeds in Germany, while in some of the European countries it is used for flavouring cheese. This seed spice is an essential component of varied cuisines such as Iranian, Mexican, Turkish, Cuban, Indian, South East Asian and Egyptian. Cumin is a regular feature in confectionary, beverages, medicines, liquors, sausages, meat, perfumery and bread manufacturing. In food processing industries cumin is used as a preservative.

Cumin oil has a characteristic flavour and odour due to the presence of falvour component called cuminaldehyde. This seed spice possesses several medicinal properties [2] such as stomachic, carminative, antimicrobial, stimulant, and astringent properties. It is widely used traditionally for treatment of various ailments cold, fever, insomnia, flatulence, diarrhoea, and other digestive disorders [3, 4]. In Indian veterinary practice cumin is popularly used as a carminative. In lactating mothers this spice is said to enhance the secretion of milk after child birth. External application of mixture of powdered cumin, honey, salt and butter is a home remedy for scorpion bite. After essential oil extraction from cumin the by-product known as *Ark Jeera* has medicinal properties which aids in indigestion. The essential oil of cumin is known to have positive effects on nervous system, reproductive system, gastrointestinal system, and immune system. Apart from this the cumin essential oil also possesses chemoprotective, antimicrobial, hypoglycaemic, antioxidant, and hypolipidemic activity but therapeutic role of secondary metabolites remains unknown. Alcohol and water extracts of cumin have nutraceutical properties such as anti-platelet aggregation, antioxidant, anti-allergic, and hypoglycemic (**Table 1**).

Sr. No	Contents	mg/100 g
1	Moisture	8.1
2	Protein	17.8
3	Fat	22.3
4	Carbohydrate	44.2
5	Fiber	10.5
6	Calcium	931
7	Potassium	1788
8	Phosphorus	449
9	Sodium	168
10	Magnesium	366
11	Vitamin B1	0.73
12	Vitamin B2	0.38
13	Vitamin C	17.20
14	Vitamin A	175 IU
15	Niacin	2.5

Sources: [5, 6]

Table 1.
Nutritional composition of cumin seeds.

1.1 Climate

Cumin grows well in tropical and subtropical areas of world ideally situated within the 20° to 38° north latitudes. Its cultivation is done best under moderately cold dry climate. Humidity during flowering and seed setting is deleterious as the crop is susceptible to fungal diseases [7]. It is requires temperature range from 9°C to 26°C for proper growth and development. Frost at the time of flowering and early fruit sett has an adverse effect on the crop. Overall, locations having low atmospheric humidity and mild winters are most suited for cumin cultivation. Frequent rains at the time of flowering, fruit setting, and maturity leads to incidence of blight and powdery mildew in the crop.

1.2 Soil

Although cumin cultivation can be done on all type of soils, most suitable are sandy soils with low organic matter, and clay or clay loam with fair organic matter.

Variety		Description
Developed by Sri Karan Narendra Agriculture University, Jobner		
RZ-19	:	Important characteristics: Erect plants with bold seeds Time to maturity: 120–140 days Average yield: 5–6 q ha^{-1}
RZ-209	:	Important characteristics: Wilt tolerance Time to maturity: 140–150 days Average yield: 6.5 q ha^{-1}
RZ-223	:	Important characteristics: Wilt tolerance, essential oil content 3.2% Average yield: 6.0 q ha^{-1}
RZ-341	:	Important characteristics: Bushy plant, semi erect growth, long and bold seeds, and tolerance to wilt, blight and powdery mildew Time to maturity: 120–130 days Average yield: 4.5 q ha^{-1}
RZ-345	:	Important characteristics: Semi erect bushy plants, long and bold seeds, and tolerance to wilt, blight and powdery mildew Time to maturity: 120–130 days Average yield: 6.07 q ha^{-1}
Developed by Sardar Krushinagar Dantiwada Agricultural University, Spice Research Centre, Jagudan		
GC-1	:	Important characteristics: Erect plants, bold seeds, and wilt tolerance Time to maturity: 105–110 days Average yield: 7.0 q ha^{-1}
GC-2:	:	Important characteristics: Bushy plants, profuse branching and attractive seeds Time to maturity: 100 days Average yield: 7.0 q ha^{-1}
GC-3	:	Important characteristics: Wilt resistant, essential oil content 3.5% Time to maturity: 100 days Average yield: 7.0 ha^{-1}
GC-4	:	Important characteristics: Resistant to Fusarium wilt Average yield: 8.75 q ha^{-1}
GC-5	:	Important characteristics: Early maturing Time to maturity: 92 days Average yield: 6.86 q ha^{-1}

Table 2.
The details of varieties.

Soil drainage is very crucial as water stagnation and excessive moisture are harmful for the crop. High soil pH also has adverse effect on the crop and preferred range of 6.8–8.3 [8], while soil suspension EC of 14 dSm^{-1} is suitable for cumin cultivation. Saline soil or saline irrigation water results in better seed filling in cumin. Apart from deep, and shallow soil, gravely soils with good porosity and drainage are also suitable for this crop.

1.3 Improved varieties

Listed below are some of the important improved varieties in India which are recommended for main growing states i.e., Rajasthan and Gujrat (**Table 2**).

2. Cultural practices

2.1 Nutrient management

Soil analysis based fertilizer application is best for obtaining good growth of crop along with optimum use of fertilizers. For a good soil structure and soil properties, 10 t ha^{-1} FYM or 5 t ha^{-1} compost can be applied three weeks prior to sowing. Using the organic wastes such as FYM provides us with an option of climate resilient crop management module while reducing the chemical fertilizer load [9]. Biofertilizers can also be used as they enhance biological activities of useful microbes in the soil along with improving the crop yield and quality [10]. As the crop responds well to fertilizer, basal application of 15 kg N, 20 kg P, and 20 kg K_2O ha^{-1} can be done, and remaining 30 kg N can be applied as top dressing 60 days after sowing.

2.2 Land preparation

Well prepared land aids in good germination and growth of the plant. Bring the soil to a fine tilth through 2–3 harrow ploughings and then level the field with a plank. At the time of third ploughing, incorporate FYM/compost in the soil. The beds of convenient size should be prepared with adequate placement of irrigation channels.

2.3 Sowing time

Time of sowing while having no influence of cost of production, has immense influence on disease and pest incidence. Therefore, it is crucial to complete sowing at appropriate time so that flowering stage escapes the period of high atmospheric humidity. For better germination, optimum temperature required is 30°C. In order to achieve good and healthy growth of the crop sowing can be done between mid-November and first week of December.

2.4 Seed rate

Optimum seed rate is essential in order to ensure ideal plant population which allows proper growth of plants leading to higher yield. Seed rate is determined by the type of variety and sowing method. Optimal seed rate ranges from 10–20 kg per hectare, where bold seeded varieties need higher seed rate.

2.5 Seed treatment

In order to control seed borne diseases, treat the seed with *Trichoderma* culture (10 g/kg seed) or alternatively with Thiram/carbendazim @ 2.5 g/kg. Inoculation with *Azospirillum* or *Azotobactor* and seed priming for 8 hours and drying of seed in shade is known to enhance the germination [11]. While, Lal et al. [12] reported improved germination after seed treatment with Bavistin @2.5 g/kg, and *Trichoderma viride* @4 g/kg.

2.6 Method of sowing

Cumin can be sown by two methods, *viz.* in lines or by broadcasting the seed, where former requires a drill while latter can be done by manual labour. Conventionally farmers go for sowing through broadcasting as it is an easier and cheaper alternative. On the other hand sowing in lines facilitated intercultural operations at all the stages of the crop. For line sowing the spacing between lines should be 25 cm, with seed sown of 1.5 cm deep covered with thin layer of soil. In case of broadcasting, cover the seed with light layer of soil with the help of a teeth rake. Avoid sowing the seed too deep, to prevent germination problems.

2.7 Crop geometry

Crop geometry is an important aspect which influences proper sunlight interception which leads to optimum physiological activities of plants. Optimal plant population aids in proper translocation of photosynthates in the plant. In case of higher than optimum plant population, competition for water, space, light, and nutrient increases which results in reduced accumulation of dry matter in plants. Therefore, optimum plant geometry is an important requirement to realise higher production of crops [13, 14]. Crop geometry in case of cumin within row spacing of 22.5–30 cm, within plant spacing of 15–30 cm.

2.8 Irrigation

Right amount of irrigation at appropriate time is essential for good germination and growth of the crop. For good germination of crop, a light irrigation should be given after sowing. Avoid heavy irrigation at this stage to avoid dislocation of the seed. After 10–12 days of sowing germination becomes visible, typically after second irrigation. In case of high temperature during day and dry weather another irrigation can also be given after an interval of 4–5 days. As per the prevailing weather and soil type irrigation is given at an interval of about 30 days. However, at the maturity stage avoid irrigation so as to prevent adverse effects on quality of seed. Sprinkler irrigation is also good in case the crop shows symptoms of wilting [15, 16]. Alternatively, drip may also be used but it is a very costly set up.

2.9 Intercultural operations

Weeds are an important problem in cumin cultivation as they compete with the crop for resources while contaminating the seed. So, in order to achieve a robust growth of the crop weed management should be done at appropriate times. First weeding and hoeing should be done when the crop has attained 4–5 cm height which happens at 30–40 days after sowing. To keep the crop free of weeds and to break the soil crust, another 2–3 weedings are needed. Some of the common weeds infesting cumin fields are zeeri (*Plantago pumila*), piazi (*Asphodellus tenifolius*), bathua (*Chenopodium album*), pili sanjee (*Melilotus indica*), khartua (*Chenopodium murale*) [17]. Among these weeds, zeeri is of serious concern as it bears close resemblance to cumin plant, and can easily become a contaminant thus reducing its crop's market value [18]. Pre-emergence application of Fluchloralin (0.77–1.00 kg) or Basalin (2.5 kg) or Stamp F-34 (3.33 kg) per hectare can be done. Besides this, pre-emergence application of Oxidiagal @75 g ha^{-1} has been found to yield very effective results for weed control at NRCSS, Ajmer.

2.10 Harvesting

Cumin generally attains maturity in about 90–120 days. Physiological maturity as indicated by yellowing of plant is the most appropriate stage to obtain produce of high quality. Complete drying of plants has adverse effect on the crop quality. Apart from this prolonged sun drying of the crop prior to threshing is not advisable as it lowers the crop quality and has deleterious effect on oil content of seed. Harvest the crop early in morning to avoid grain shattering, followed by beating and trampling it on a clean threshing floor. In case of large scale cultivation threshing can also be with the help of a thresher which is very popular among the farmers now days. There have been attempts to design modified seed harvesters for cumin [19]. After harvesting cumin is cleaned and graded with the help of mechanical devices. The thoroughly cleaned cumin seeds can be stored with 7–8% moisture level at 40% equilibrium relative humidity. Seeds can be stored in polythene film lined gunny bags in a well ventilated, dry and cool place till next sowing season.

3. Post-harvest technology

After harvest cumin seeds are dried under partial shade in order to keep the moisture up to 9% for storage. Higher moisture content in seeds increases the chances of fungal contamination during storage. The cleaned, dried, and graded seed is packed in the standard size packaging with appropriate labelling. Storage of dried seeds is done in environment friendly plastic film lined gunny bags. While processing cumin, care should be taken to maintain the vitality of the organic ingredient. Selection of processing method should be in such a manner that number and quantity of additives as well as processing aids is kept to a minimum. Essential oil extraction is done by distillation of mature dried seeds, and generally it is done by hydro or steam distillation. For oil extraction, dried seeds can be crushed or steam distilled wholly to yield 2.5 to 4.5% oil content. In International market, there is a good demand for oleoresin obtained from cumin. After processing, various products are produced such as cumin powder, essential oil, cumin oleoresin and fixed oil. Volatile oil after extraction should be stored in well sealed containers.

3.1 Crop protection

Crop protection covers all biotic and abiotic aspects damaging crop. In the cumin growing regions frost is the major concern affecting the production. Cumin is mainly susceptible to frost which occurs from flowering to seed development stage of crop. It can be managed by proper irrigation during frost incidence with smoke or wind breaker.

3.2 Biotic stress

Blight, wilting and powdery mildew are three main diseases which hamper cumin production.

Disease	Symptoms	Control measures
Blight:	Symptoms appear mainly on leaves and stem in form of dark brown lesions which results in withering of plants	• Spray of Mancozeb (0.2%) or Azoxystrobin (0.1%) or Difenoconazole (0.05%) in 500 l water per hectare at disease appearance. Repeat after 15 days if needed. • Spray scheduling: At 45–60 days of sowing spray Mancozeb, followed by 2–3 need based sprays of difeno-conazole or azoxystrobin at fortnightly interval.
Wilt	Wilt infection ususally occurs in patches and leads to drooping of leave and epinasty. Disease is very difficult to manage and severly infected plants usually wither and die	• Ploughing during summer season to sterlize soil • 3 year crop rotation with non-host plants • Disease free seed • Application of *Trichoderma* species @ 10 g/kg for seed treatment and for soil application mix 2.5 kg of consortia 50 kg FYM per hectare
Powdery mildew	In initial infection, leaves and twigs show appearance of white powdery spots, and upon severe infection the powdery mass covers whole plant	• Sulphur dusting @ 20–25 kg/ha or foliar spray of Wettable Sulphur 0.2% at fortnightly interval commenc-ing from disease appearance in the field

3.3 Insect pests

Insect-pest	Damage	Management
Aphid (*Aphids gossypii* and *Myzus persicae*)	Aphids affect the crop by sucking sap of the plant. Populations build up initiates at vegetative stage and peak build up occurs during flowering (February) to seed formation. In case of unprotected crops, more than 50% yield loss can occur.	• For monitoring and trapping use yellow sticky traps • Timely sowing of crop in desired geometry. • Application of recommended nitrogenous fertilizer doses as higher dose increases succulency. • Foliar spray of Neem oil (2%) or Neem Seed Kernel Extract (5%) • Spray of *Verticillium lecanii* (108 spores/g) at the dose of 5.0 g/lire of water • For severe infestations spray Metasytox (0.03%) or Dimethoate (0.03%) or Imidacloprid (0.005%) or Emamectin benzoate (10 g ai/ha)
Thrip (*Thrips tabaci*)	Attack of the insect is first visible during early vegetative growth of crop and continues up to flowering stage. Thrips damage the crop by sucking the sap from plant which ultimately results in leaf yellowing and drying. Higher population resulted drying of whole plants.	• Use of botanicals and chemical pesticides same as that for aphids. Besides that Thiomethoxam (0.025%) has also been recommended for control of thrips.

4. Conclusion

Cumin finds wide use in kitchen as well as in traditional medicine. It a nutritionally rich crop, with a distinct flavour as a virtue of the essential oil present in it. Flavour of cumin is direct function of its essential oil content. Also in recent times, popularity of cumin along with other spices and condiments have risen, which has led to increase in its demand. India the global leader in production of cumin, and a major consumer of this produce. All these factors make cumin an active cash crop which can be explored further in areas where it is not grown presently. So, we have outlined the successful cultivation operations which are practised in the main cumin growing areas of India. Adoption of scientific package of practices for cultivation will boost the production, while dealing with the constraints in an effective way.

Author details

Navjot Singh Brar[1], Prakash Mahala[2], Kartik Sharma[3], Parmdeep Singh Dhanda[4], Alpa Yadav[5], Meenakshi Sharma[6] and Prashant Kaushik[7]*

1 Vegetable Research Station, Khanaura, Hoshaipur, India

2 Regional Research Station, Abohar, Fazilka, India

3 Department of Agronomy, Punjab Agricultural University, Ludhiana, Punjab, India

4 Department of Biochemistry, Punjab Agricultural University, Ludhiana, Punjab, India

5 Department of Botany, Indra Gandhi University, Meerpur, Rewari, India

6 Department of Chemistry, Ranchi University, Ranchi, India

7 Instituto de Conservación y Mejora de la Agrodiversidad Valenciana, Universitat Politècnica de València, Valencia, Spain

*Address all correspondence to: prakau@doctor.upv.es

References

[1] Dar EA, Mehdi M, Ahmad M, Bhat FN, Hussain N, Hussain M, et al. Cumin: The flavour of Indian cuisines-history, cultivation and uses. Chemical Science Review and Letters. 2019;**8**:129-135

[2] Singh RP, Gangadarappa HV, Mruthunjaya K. *Cuminum cyminum* – A popular spice: An updated review. Pharmacognosy Journal. 2017;**9**:1-5

[3] De M, De AK, Mukhopadhvay R, Banerjee AB, Micro M. Antimicrobial activity of *Cuminum cyminum* L. ARS Pharmaceutica. 2003;**44**:257-269

[4] Malhotra SK, Vashishtha BB. Package of Practices for Production of Seed Spices. Ajmer: Book, Publisher Director, National Research Centre on Seed Spices; 2008. p. 40

[5] Parthasarathy VA, Kandiannan K. Horticulture: Vegetable Science, Spices and Condiments. New Delhi, India: Indian Institute of Spices Research (ICAR); 2007

[6] Kakhki AH, Mohassel MS. Processing, Chemical composition and standards of cumin. In: Kafi M, Mohassel MHR, Koocheki A, Nassiri M, editors. Cumin Production and Processing. Enfield, USA: Science Publishers; 2006

[7] Kafi M. Historical background, regions of production and application of cumin. In: Kafi M, Mohassel MHR, Koocheki A, Nassiri M, editors. Cumin Production and Processing. Enfield, USA: Science Publishers; 2006

[8] Lal G. Scenario, importance and prospects of seed spices: A review. Current Investigation in Agriculture Research. 2018;**8**:49-62

[9] Kumar A, Prakash CHB, Brar NS, Kumar B. potential of vermicompost for sustainable crop production and soil health improvement in different cropping systems. International Journal of Current Microbiology and Applied Sciences. 2018;**7**:1042-1055

[10] Brar NS, Thakur KS, Kumar R, Mehta DK, Sharma N, Kumar D, et al. Effect of organic manures and biofertilizers on fruit yield and its contributing traits of tomato (*Solanum lycopersicum* L.). Ecology Environment and Conservation. 2015;**21**:1783-1787

[11] Malhotra SK, Vashishtha BB. Possibilities of mechanization in seed spices. In: Production, Development, Quality and Export of Seed Spices. New Delhi. 2010. pp. 28-30

[12] Lal G, Saran PL, Devi G, Deepak RR. Production technology of cumin (*Cuminum cyminum* L.). Advances in Vegetable and Agronomy. 2014;**1**:223-231

[13] Kamboj N, Batra VK, Brar NS, Rana MK. Effect of various plant density at different levels of phosphorous and potash on growth and seed yield of onion (*Allium cepa* L.) cv. Hisar-2. Indian Journal of Agricultural Research. 2017;**51**:514-517. DOI: 10.18805/IJARe.A-4749

[14] Mahala P, Jaidka M, Sharma M, Brar NS. Effect of crop geometry on growth and yield of kharif onion. Journal of Krishi Vigyan. 2019;7:267-269

[15] Ravindran PN, Babu NK, Shiva KN, Kallupurackal JA. Advances in Spices Research History of Achievements of Spices Research in India Since Independence. Jodhpur, Rajasthan; 2006. p. 994

[16] Sundria MM, Mehriya ML, Rathore BS, Choudhary BR. Cumin (*Cuminum cyminum* L.) sustainable production technology in Rajasthan. Jodhpur, Rajasthan: Agricultural University; 2014. p. 29

[17] Anonymous. Integrated Pest Management Package for Cumin. India: Ministry of Agriculture; 2002. p. 11

[18] Kumar M, Sahoo PK, Kushwaha DK, Saxena SN, Mani I, Singh G, et al. Cumin cultivation: Present status and future prospects. The Pharma Innovation Journal. 2021;**10**:1121-1123

[19] Rahimi M. Chemical Control of Weeds in Cumin. Khorasan, Iran: Iranian Scientific and Industrial Research Organization, Khorasan Center; 1993

Chapter 11

Essential Oil of Ginger: Effect of Cultivation and Uses

Temitayo Esther AdeyeOluwa

Abstract

Ginger (*Zingiber officinale* Rosc.) is a spice used in many parts of the world for culinary and medicinal purposes. It is a good source of essential oil with both the rhizome and its essential oil becoming increasingly acceptable for traditional, medicinal and commercial uses. Essential oils may be referred to as ethereal oils or volatile oils due to their volatile nature at room temperature. This review is intended to highlight the uses of ginger essential oil as well as summarise the effect of site, duration and geographical location of cultivation on the oil. In view, there are vast and abundant uses of ginger essential oil and different cultivars of ginger would be observed to differ in weight yield and composition, with China ginger oil (4.07% yield) having 43 compounds and Indian ginger oil (1.26% yeild) having 60 compounds, hence differing in quality and bioactivity. It may be concluded in this review that various aspects of cultivation as earlier mentioned affect the composition, bioactivity, potency, colour, aroma and weight yield of ginger essential oil which essentially affect its use from one culture to another.

Keywords: ginger, essential oil, uses, cultivation, composition

1. Introduction

Ginger is a widely cultivated food and medicinal crop in many parts of the world. It is amongst the most widely used spices globally [1]. It was first cultivated in Ethiopia in the 1200s and continues to be cultivated despite challenges of low quality, reduced economic output and poor handling by farmers and sellers [2]. Although with similar health benefits as fresh ginger, ginger essential oil is said to be the most potent portion of ginger [3].

The essential oil of ginger is the yellowish or green coloured volatile oil extracted from ginger rhizomes, which is about 1–3% of a rhizome. It may be obtained from fresh ginger rhizome, dried rhizomes or ginger peel, with oil from the fresh ginger reported to be of better fragrance than that from dried ginger [4, 5]. As it is known with spices, ginger oil is generally regarded as safe [GRAS] [6, 7]. It is known as the oil of empowerment because it gives the user a feeling of self-assurance and courage [3].

Ginger oil may be extracted from ginger rhizomes in different ways. It can be obtained traditionally (when it is crushed, macerated in water for about 12 h, heated with sesame oil, filtered and centrifuged), conventionally (crushed, macerated in sesame oil for 7 days, filtered and centrifuged) or by hydro distillation (with the use of

Clevenger-type apparatus) [8]. GC–MS analysis of ginger essential oils shows its major constituents to be Zingiberene, β-bisabolene, ar- curcumene, camphene, citral and geranial [5].

The volatile oil of ginger possesses extensive biological effects. The essential oil of ginger has been indicated as an antispasmodic, anti-inflammatory, antinociceptive and antioxidant agent [9–12]. It has also been used in the treatment of joint stiffness and pain [13] as well as a tonic for the uterus, brain and the stomach [14]. In addition, it is useful in managing respiratory disorders [15], relieves anxiety and is heart and liver friendly [16–18]. In fact, it has been proffered that the benefits derived from the use of essential oils of ginger is similar to that derived from fresh ginger [3], suggesting that the chemical constituents of fresh ginger are retained in its oils. It possesses antiseptic, antimicrobial, healing, anti-nausea and anti-stress properties and is thus employed in facilitating healing after surgical procedures [19, 20]. It aids digestion and is used in the treatment of stomach disturbances [21]. In addition, the essential oil of ginger is useful as a flavouring, preservative, perfumery and pesticidal agent [22]. These properties have been attributed to the volatile constituents such as monoterpenes and sesquiterpenes and pungent fractions of ginger, specifically, shogaols, gingerols and paradols [23]. The constituents of ginger essential oil are however affected in composition, potency and quantity by place, duration and region of cultivation, thus affecting its bioactivity, colour, aroma and ultimately its use.

2. Uses of ginger essential oil

2.1 As spice

Globally, ginger is used as spice in foods in order to add flavour and nutrient to food. In many cultures, it is used in the making of several dishes, adding to them a hot and distinctive spicy flavour [24]. It may be used fresh or in the dried form and is an essential ingredient of most spice blends [25]. The inclusion of ginger oil in food has been demonstrated to improve the quality and flavour of the food [26].

2.2 As a herbal drug

In ayurveda, Chinese or Tibb-Unani traditional medicine, ginger essential oil has been extensively employed [5, 27]. Indeed, spices from the family Zingiberacea, including ginger, are well utilised in traditional medicine for the management and treatment of several conditions and diseases such as stomach ailments and indigestion, respiratory disorders, infections and anxiety [3, 11]. In Iranian traditional medicine, it is believed that traditional medicine is more potent when formulated with ginger oil, as against when ginger rhizome is used in its formulation [8]. In Indian traditional medicine, it is used as stop excessive blood clotting while in Arabian traditional medicine, it is used as an aphrodisiac [28].

2.3 Antimicrobial agent

The antimicrobial activity of ginger essential oil has been linked with the presence of abundant oxygenated compounds present [5]. At 12% V/V, it was found to be effective in reducing *Listeria monocytogenes* burden in food [1]. Its antifungal activity has also been demonstrated by [29]. Ginger essential oil may therefore act as a preservative agent in food and its inclusion or marination into food may hence enhance the shelf life of food [26]. This has been found to be better achieved with the volatile oil encapsulated in protein fibres [30], thus enabling a gradual and controlled release of the oil into the food item, enhancement of solubility, as well as prevention of alteration in the food taste and thermal destruction of the oil [31].

2.4 As a supplement

Dietary supplementation with essential oil of ginger has been recorded to improve immunological, haematological and biochemical parameters as well as performance of broilers [32]. This is proposed to enhance organic poultry which is evidently healthier and was recently recommended in the poultry industry. Supplementing fish diet with ginger oil also improved resistance to disease, immune response and growth of the fishes [33]. In addition, ginger supplement in broiler feed increased feed conversion rate, weight gain, improved blood parameters and general performance of the birds [34]. Elazab et al. [35] also reported how the inclusion of ginger essential oil in rabbit diet improved meat quality, growth performance and weight gain.

2.5 As an antioxidant

The essential oil of ginger is a potent antioxidant agent [36]. It was reported by [37] to improve the antioxidant status of heat stressed broilers and has been proposed to be a better source of antioxidants in the diet of broilers than synthetic antioxidants. In addition, ginger essential oil inhibits *in vitro* lipid peroxidation as well as scavenging hydroxyl and superoxide radicals [11]. Antioxidant enzymes, such as glutathione (GSH), superoxide dismutase (SOD), catalase and glutathione peroxidase (GPx), which were reduced in oxidative stressed rats was reported to increase with the oral administration of ginger essential oil [21].

2.6 Anti-cancer agent

Bioactive compounds present in ginger such as 6-gingerol have been documented to bring about apoptosis in cancerous cells. Treatment with the essential oil of ginger was seen to reduce tumour sizes by over 50% [11].

2.7 Anti-inflammatory agent

The use of ginger oil in the reduction of inflammation has been documented by [11]. The aromatic components of the essential oil of ginger such as terpenes are responsible for its anti-inflammatory activity [38]. Ginger oil also alters T cell non-specific production and cell-mediated immune response [39].

2.8 For commercial use

Ginger oil was largely produced in Europe and America but is now a great commercial product globally especially in Jamaica, China, Indonesia, India and Australia [4]. Ginger essential oil is utilised in production companies as a preservative (an antimicrobial), a food enhancer (to improve taste or flavour) or an additive. It has been employed in the production of bakery products, pharmaceuticals, beverages, pesticides, hair care products, aromatic oils for aromatherapy amongst others.

2.9 As an antinausea, antiemetic and gastroprotective agent

The antinausea, antiemetic and gastro-protective activities of ginger has been extensively documented [40]. It has been said that the primary use of ginger is to combat nausea [28]. A study by [41] showed that the inhalation of the volatile oil of ginger reduced the post-operative occurrence of nausea and vomiting in human subjects that underwent abdominal surgery. Ginger essential oil was also shown to inhibit gastric ulcer in rats by 85.1% which was confirmed by histopathology of the stomach lining [21].

2.10 As beverage

There is an increased cognizance on the health benefits of consuming ginger-based beverages thereby, leading to growth in its demand [42]. The use of natural products such as ginger to spice up beverages is currently more acceptable than the use of chemicals or synthetic agents, as natural products are known to enhance health by enriching food products with bioactive phytoconstituents and other essential nutrients needed by the body [43].

2.11 As perfume

The strong aroma of ginger has made it a very good source of perfume [29]. This aroma has been attributed to gingerol, a pungent ketone, present in large quantity in ginger oil. Aromatherapy has been employed in the curation of nausea and vomiting in breast cancer patients exposed to chemotherapy. Thus, inhalation of ginger oil has been ascertained to be an effective complementary therapy in most cases of nausea and vomiting induced by chemotherapy [44].

2.12 Other uses

Other uses of ginger essential oil include its antispasmodic, antidiabetic, antihypertensive, neuroprotective, antiplatelet aggregation, anticholesterolemic, antiseptic, carminative, appetite enhancer and expectorant uses [40]. It also promotes sweating, soothes aches, relaxes peripheral blood vessels as well as stimulates blood circulation [45].

3. Effect of cultivation

Effective systemic cultivation of essential oil producing plants is vital for the constant production of essential oil. Many factors influence the production of essential oil producing plants and may thus affect the production of the oils themselves [46]. Improved cultivars are now seen to give better quality and quantity of oil [5].

3.1 Chemical composition

Cultivation site as well as duration of cultivation has been said to affect the chemical composition of the essential oil of ginger, as well as their profile/constituents. The geographical location of the world in which ginger is cultivated also affects its chemical composition [47]. For instance, ginger grown in China was found with 43 different compounds [48] while the same species of ginger grown in Ghaziabad (India) possessed

80 compounds with varied proportions reported [49]. At Fiji, the compounds present in the same species of ginger were varied from that in India [47]. This variation in chemical composition has affected the therapeutic and commercial use of ginger essential oil [5]. An instance would be the reported proportion of zingiberene which was found to be varied depending on the country of cultivation. 9.5% was found in ginger cultivated in Mauritius while 29.0% was found in Nigerian ginger, 1.3% in Sri Lankan ginger, 38.12% in Chinese ginger, 20–28% in Australian ginger, 46.2% in Indian ginger and so on [4]. The variation in the chemical composition of 17 cultivars of ginger from North India has also been documented (**Table 1**) [58].

3.2 Bioactivity

Since the chemical composition of essential oil is seen to be affected by place and/or duration of cultivation, it may be proposed that the cultivation also affects the bioactivity of the essential oil of ginger [5]. It is expected that since cultivation alters the chemical composition and peak of the different compounds present in ginger, the activity of a specific volatile oil of ginger will be dependent on the compounds present in the oil, its quantity as well as the intensity or peak of the compounds. For instance, sesquiterpenes which are known to be effective antioxidant and anti-inflammatory agents [59] are not abundant in Sri Lankan ginger essential oil while monoterpenes which are strong antimicrobial agents [60] are abundant in it [47].

3.3 Potency

The peak intensity of ginger essential oil cultivated in different areas was considered and it was found that both the site and duration of cultivation affected the quality

Region of cultivation	Weight yield	Number of compounds	Colour	Dominant compound
China	4.07% [48]	43 [48]	Pale yellow to Amber [7]	38.12% Zingiberene [4]
Ghaziabad (India)	1.26% [49]	80 [46]	Yellow [7]	46.2% Zingiberene [4]
West Africa (Nigeria)	1.02%-2.4% [7]	54 [50]	Pale yellow [7]	29% Zingiberene (Nigeria) [4]
Australia	NA	NA	Yellow [51]	Citral (Neral and geraniol). 20-28% Zingiberene [4]
Sri Lanka	NA	NA	Golden [52]	Antimicrobial compounds (Monoterpenes). 1.3% Zingiberene [4]
Iran	1.2% [53]	55-59 [8]	NA	β-sesquiphellandrene, Zingiberene, caryophyllene, ar-curcumene, farnesene [29]
Ecuador	NA	70 [54]	NA	α-zingiberene (17.4%) [54]
Brazil (Rio-de-Janeiro)	2.3% [55]	NA	NA	α-Zingiberene [56]
Veitnam	2.1% [57]	NA	NA	Ar-curcumene [57]

Table 1.
Effect of region of cultivation on ginger essential oil.

or potency of compounds present in the essential oil [47]. The variation in the degree of pharmacological activity of the essential oil of ginger from different cultivars has been documented. It was noted that the intensity of the different compounds in different cultivars affects the potency of ginger essential oil. Essential oil of ginger from Brazil, for instance, was found to be more effective against *Staphylococcus aureus* bacteria compared with ginger essential oil from Veitnam. This has been attributed to variation in quantity of ar-curcumene, α-zingiberene, geranial, α-farnesene, β-bisabolene and β-sesquiphellandrene in both cultivars (**Table 1**) [7].

3.4 Aroma and commercial use

Australian ginger essential oil has a "lemony" odour due to the presence of high quantity (about 51–71%) of citral (neral and geranial). Jamaican cultivar contains higher levels of sesquiterpenes and lower citral content than the Australian cultivar. This is responsible for the intensity of aroma and pungency detected in the ginger oil with higher proportions of gingerols and other pungent compounds, thereby probably making it more commercially viable than those with less aroma (**Table 1**) [61].

3.5 Weight yield

The weight yield of ginger essential oil is also affected by cultivation. Kiran et al. [58] reported that the period of cultivation affected the weight yield of essential oil. In that study, it was found that ginger grown for only 5 months produced more essential oil than those grown for 7–8 months. This study was carried out in Sri Lanka and it is proposed that the higher essential oil yield in the less mature ginger is due to less fibre content and lower chemical intensity of the rhizome [58]. The variation in the quantity of essential oils produced from 17 cultivars of ginger from North India has been documented [58]. Indeed, different cultivars of ginger have been said to possess varied oil yielding capacities. This was reiterated by [5] who reported a difference in the proportion of ginger essential oil from different cultivars. In other studies, ginger grown in China was reported to give a 4.07% yield of the total weight of ginger [48] while that grown in India (Ghaziabad) yielded 1.26% of total weight of ginger. This disparity in quantity has affected the commercial availability of ginger essential oil significantly (**Table 1**) [5].

3.6 Colour

It has been reported that ginger essential oil derived from different geographical regions of the world differ in colour. Essential oil of ginger from West Africa is pale yellow, from India, it is yellow while that from China is pale yellow to amber. This might also affect acceptability and hence its commercial use. The colour variation has been attributed to alterations in the chemical composition of the oil from one region to another (**Table 1**) [7, 48].

4. Conclusion

The numerous uses for ginger and its essential oil continues to draw the attention of many to it. However, careful considerations must be taken in the selection of ginger essential oil due the effect of cultivation on the oil. The composition, potency, bioactivity in addition to colour, aroma and quality of this essential oil is altered from one region of the world to another. This variation is also observed with the duration as well as the season of cultivation. As the site, duration, period or season of cultivation of ginger may influence the presence and quantity of specific compounds, its chemical activity, the weight yield as well as the potency, odour, and taste of the oil, oils derived from different regions of the world should be analysed for their chemical components and properly documented in order that maximal benefits may be derived by the user.

Acknowledgements

The author acknowledges the management of Intechopen for the opportunity given to publish in this book.

Conflict of interest

The author declares no conflict of interest.

Author details

Temitayo Esther AdeyeOluwa
University of Ibadan, Ibadan, Nigeria

*Address all correspondence to: temitayoadeyeoluwa@gmail.com

References

[1] da Silva FT, da Cunha KF, Fonseca LM, Antunes MD, El Halal SLM, Fiorentini ÂM, et al. Action of ginger essential oil (*Zingiber officinale*) encapsulated in proteins ultrafine fibers on the antimicrobial control in situ. International Journal of Biological Macromolecules. 2018;**118A**:107-115. DOI: 10.1016/j.ijbiomac.2018.06.079

[2] Geta E, Kifle A. Production, processing and marketing of ginger in Southern Ethiopia. Journal of Horticulture and Forestry. 2011;**3**: 207-213 Corpus ID: 131469848

[3] Ugbabe GE, Okhale SE, Ashwe JD, Egharevba HO, Ibrahim JA, Kunle OF. Comparative studies of essential oils from Zingiber officinale grown in Nigeria. Journal of Phytomedicine and Therapeutics. 2019;**18**(1):237-252. eISSN: 1118-1028

[4] Bag BB. Ginger processing in India (Zingiber officinale): A review. International Journal of Current Microbiology and Applied Science. 2018; 7(04):1639-1651. DOI: 10.20546/ijcmas.2018.704.185

[5] Munda S, Dutta S, Haldar S, Lal M. Chemical analysis and therapeutic uses of ginger (*Zingiber officinale* Rosc.) essential oil: A review. Journal of Essential Oil Bearing Plants. 2018;**21**(4): 994-1002. DOI: 10.1080/0972060X.2018.1524794

[6] Govindarajan VS. Ginger- chemistry, technology and quality evaluation: Part 1. Critical Reviews in Food Science and Nutrition. 1982;**17**(1):1-96. DOI: 10.1080/10408398209527343

[7] Mahboubi M. Zingiber officinale Rosc. essential oil, a review on its composition and bioactivity. Clinical Phytoscience. 2019;**5**(6):1-12. DOI: 10.1186/s40816-018-0097-4

[8] Shirooye P, Mokaberinejad R, Ara L, Hamzeloo-Moghadam M. Volatile constituents of ginger oil prepared according to Iranian traditional medicine and conventional method: A comparative study. African Journal of Traditional, Complementary and Alternative Medicine. 2016;**13**(6):68-73. DOI: 10.21010/ajtcam.v13i6.11

[9] Yassin N, Elrokh E, Elshenawy S, Ibrahim B. The study of the antispasmodic effect of Ginger (*Zingiber officinale*) *in vitro*. Der Pharmacia Lettre. 2012;**4**:263-274. ISSN 0974-248X

[10] Mashhadi NS, Ghiasvand R, Askari G, Hariri M, Darvishi L, Mofid MR. Anti-oxidative and anti-inflammatory effects of ginger in health and physical activity: review of current evidence. International Journal of. Preventive Medicine. 2013;**1**:S36-S42. PMID: 23717767; PMCID: PMC3665023

[11] Jeena K, Liju VB, Kuttan R. Antioxidant, anti-inflammatory and antinociceptive activities of essential oil from ginger. Indian Journal of Physiology and Pharmacology. 2013;**57**: 51-62. PMID: 24020099

[12] Akinyemi JA, Adeniyi PA. Effect of essential oils from ginger (*Zingiber officinale*) and turmeric (*Curcuma longa*). Rhizomes on some inflammatory biomarkers in cadmium induced neurotoxicity in rats. Journal of Toxicology. 2018;**2**:1-7. DOI: 10.1155/2018/4109491

[13] Sritoomma N, Moyle W, Cooke M, O'Dwyer S. The effectiveness of Swedish massage with aromatic ginger oil intreating chronic low back pain in older

adults: A randomized controlled trial. Complementary Therapies in Medicine. 2014;**22**:26-33. DOI: 10.1016/j.ctim.2013.11.002

[14] Moamen TM. Tohfe al-Momenin. Tehran: Shahid Beheshti University of Medical Sciences; 2007. p. 232

[15] Townsend EA, Matthew ES, Yi Z, Carrie X, Bhupinder H, Charles WE. Effects of ginger and its constituents on airway smooth muscle relaxation and calcium regulation. American Journal of Respiratory Cell and Molecular Biology. 2013;**48**(2):157-163. DOI: 10.1165/rcmb.2012-0231OC

[16] Samira K, Masoomeh K, Zahra BM, Hamed F, Amir K, Mani J. Effect of treatment with ginger on the severity of premenstrual syndrome symptoms. ISRN Obstetrics and Gynecology. 2014: 792708. DOI: 10.1155/2014/792708

[17] Fuhrman B, Rosenblat M, Hayek T, Coleman R, Aviram M. Ginger extract consumption reduces plasma cholesterol, inhibits LDL oxidation and attenuates development of atherosclerosis in atherosclerotic, apolipoprotein E-deficient mice. Journal of Nutrition. 2000;**130**(5): 1124-1131. DOI: 10.1093/jn/130.5.1124

[18] Liu CT, Raghu R, Lin SH, Wang SY, Kuo CH, Tseng YJ, et al. Metabolomics of ginger essential oil against alcoholic fatty liver in mice. Journal of Agricultural and Food Chemistry. 2013;**61**(46): 11231-11240. DOI: 10.1021/jf403523g

[19] Qidwai W, Alim SR, Dhanani RH, Jehangir S, Nasrullah A, Raza A. Use of folk remedies among patients in Karachi Pakistan. Journal of Ayub Medical College Abbottabad. 2003;**15**(2):31-33. PMID: 14552245

[20] Stea S, Beraudi A, De Pasquale D. Essential oils for complementary treatment of surgical patients: State of the art. Evidence Based Complementary and Alternative Medicine. 2014;726341. DOI: 10.1155/2014/726341

[21] Liju VB, Jeena K, Kuttan R. Gastro protective activity of essential oils from turmeric and ginger. Journal of Basic Clinical Physiology and Pharmacology. 2014;**26**(1):95-103. DOI: 10.1515/jbcpp2013-0165

[22] Shubha RS. Medicinal uses of Ginger to improve growth and enhance immunity in aquaculture. International Journal of Chemical Studies. 2015;**3**(2): 83-87. Corpus ID: 26465832

[23] Masood SB, Sultan MT. Ginger and its health claims: Molecular aspects. Critical Reviews in Food Science and Nutrition. 2011;**51**(5):383-393. DOI: 10.1080/10408391003624848

[24] Srinivasan K. Ginger rhizomes (Zingiber officinale): A spice with multiple health beneficial potentials. Pharma Nutrition. 2017;**5**(1):18-28. DOI: 10.1016/j.phanu.2017.01.001

[25] Kikuzaki H. Ginger for drug and spice purposes. In: Mazza G, Oomah BD, editors. Herbs, Botanicals and Teas. Functional Foods and Nutraceuticals Series. USA: CRC Press; 1998. p. 77. ISBN-13: 978-0367398521. ISBN-10: 0367398524

[26] Jakribettu RP, Boloor R, Bhat HP, Thaliath A, Haniadka R, Rai MP, et al. Ginger (*Zingiber officinale* Rosc.) oils. In: Preedy VR, editor. Essential Oils in Food Preservation, Flavor and Safety. 1st ed. Academic Press; 2016. pp. 447-454. DOI: 10.1016/B978-0-12-416641-7.00050-X. ch50

[27] El-Ghorab AH, Nauman M, Anjum FM, Hussain S, Nadeem MA. A comparative study on chemical

composition and antioxidant activity of ginger (*Zingiber officinale*) and cumin (*Cuminum cyminum*). Journal of Agricultural and Food Chemistry. 2010; **58**:8231-8237. DOI: 10.1021/jf101202

[28] Moghaddasi MS, Kashani HH. Ginger [Zingiber officinale]: A review. Journal of Medicinal Plants Research. 2011;**6**(26):4255-4258. DOI: 10.5897/JMPR11.787

[29] Noshirvani N, Ghanbarzadeh B, Gardrat C, Rezaei MR, Hashemi M, Le Coz C, et al. Cinnamon and ginger essential oils to improve antifungal, physical and mechanical properties of chitosan-carboxymethyl cellulose films. Food Hydrocolloids. 2017;**70**:36-45. DOI: 10.1016/j.foodhyd.2017.03.015

[30] Restuccia D, Spizzirri GU, Parisi OI, Cirillo G, Lemma F, Pouci F, et al. New EU regulation aspects of global market of active and intelligent packaging for food industry application. Food Control. 2010;**21**:1425-1435. DOI: 10.1016/j.foodcont.2010.04.028

[31] Antunes MDGS, Dannenberg AM, Fiorentini VZ, Pinto L-T, Lim ER, Dias Zavareze ARG. Antimicrobial electrospun ultrafine fibers from zein containing eucalyptus essential oil/cyclodextrin inclusion complex. International Journal of Biological Macromolecules. 2017;**104**:874-882. DOI: 10.1016/j.ijbiomac.2017.06.095

[32] Saleh N, Allam T, Abd El-latif A, Ghazy E. The effects of dietary supplementation of different levels of thyme (*Thymus vulgaris*) and ginger (*Zingiber officinale*) essential oils on performance, heamtological, biochemical and immunological parameters of broiler chickens. Global Veterinaria. 2014;**12**(6):736-744. DOI: 10.5829/idosi.gv.2014.12.06.83189

[33] Brum A, Pereira SA, Owatari MS, Chagas EC, Chaves FCM, Mouriño JLP, et al. Effect of dietary essential oils of clove basil and ginger on Nile tilapia (*Oreochromis niloticus*) following challenge with *Streptococcus agalactiae*. Aquaculture. 2017;**468**(1):235-243. DOI: 10.1016/j.aquaculture.2016.10.020

[34] Ifelayo II, Oyewole BO, Adesola MA, Anjorin GO. Performance and haematology of broiler strains (cobbs and arbor-acre) fed ginger-based diet at the early phase. GSC Biological and Pharmaceutical Sciences. 2020;**11**(01):197-206. DOI: 10.30574/gscbps.2020.11.1.0107

[35] Elazab MA, Khalifah AM, Elokil AA, Elkomy AE, Rabie MM, Mansour AT, et al. Effect of dietary rosemary and ginger essential oils on the growth performance, feed utilization, meat nutritive value, blood biochemicals, and redox status of growing NZW rabbits. Animals (Basel). 2022;**2**(3):375-389. DOI: 10.3390/ani12030375

[36] Pires JRA, de Souza VGL, Fernando AL. Chitosan/montmorillonite bio nanocomposites incorporated with rosemary and ginger essential oil as packaging for fresh poultry meat. Food Packaging and Shelf Life. 2018;**17**:142-149. DOI: 10.1016/j.fpsl.2018.06.011

[37] Habibi R, Sadeghi GH, Karimi A. Effect of different concentrations of ginger root powder and its essential oil on growth performance, serum metabolites and antioxidant status in broiler chicks under heat stress. British Poultry Science. 2014;**55**(2):228-237. DOI: 10.1080/00071668.2014.887830

[38] Funk JL, Frye JB, Oyarzo JN, Chen J, Zhang H, Timmermann BN. Anti-inflammatory effects of the essential oils of ginger (*Zingiber officinale* Roscoe) in experimental rheumatoid arthritis.

PharmaNutrition. 2016;**4**(3):123-131. DOI: 10.1016/j.phanu.2016.02.004

[39] Wilasrusmee C, Kittur S, Siddiqui J, Bruch D, Wilasrusmee S, Kittur DS. *In vitro* immunomodulatory effects of ten commonly used herbs on murine lymphocytes. The Journal of Alternative and Complementary Medicine. 2002; **8**(4):467-475. DOI: 10.1089/ 107555302760253667

[40] Balogun FO, AdeyeOluwa ET and Ashafa AOT. Pharmacological Potential of Ginger. In: Wang H, editor. Ginger Cultivation and its Antimicrobial and Pharmacological Potentials. London, UK: IntechOpen; 2019; DOI: 10.5772/ intechopen.88848

[41] Lee YR, Shin HS. Effectiveness of ginger essential oil on postoperative nausea and vomiting in abdominal surgery patients. The Journal of Alternative and Complementary Medicine. 2017;**23**(3):196-200. DOI: 10.1089/acm.2015.0328

[42] Gaikwad KK, Singh S, Shakya B. Studies on the development and shelf life of low calorie herbal aonla-ginger RTS beverage by using artificial sweeteners. Journal of Food Processing & Technology. 2012;**4**:2. DOI: 10.4172/ 2157-7110.1000200

[43] Nutakor C, Essiedu JA, Adadi P, Kanwugu ON. Ginger beer: An overview of health benefits and recent developments. Fermentation. 2020;**6**(4): 02. DOI: 10.3390/fermentation6040102

[44] Lua PL, Salihah N, Mazlan N. Effects of inhaled ginger aromatherapy on chemotherapy-induced nausea and vomiting and health-related quality of life in women with breast cancer. Complementary Therapies in Medicine. 2015;**23**(3):396-404. DOI: 10.1016/j. ctim.2015.03.009

[45] Murray H. Essential Oils: Art, Agriculture, Science, Industry and Entrepreneurship (a Focus on the Asian-Pacific Region). Nova Scientific Publishers; 2009. pp. 626-633. ISBN: 978-1607418658

[46] Bhattacharya S.. Cultivation of essential oils. In: Preedy VR, editor. Essential Oils in Food Preservation, Flavor and Safety. Academic Press; 2016. pp. 19-29. DOI:10.1016/B978-0-12-416641-7.00003-1.ch3

[47] Jayasundara N, Arampath P. Effect of variety, location & maturity stage at harvesting, on essential oil chemical composition, and weight yield of Zingiber officinale roscoe grown in Sri Lanka. Heliyon. 2021;**7**(3):e06560. DOI: 10.1016/j.heliyon.2021.e06560

[48] Feng J, Du Z, Zhang L, Luo W, Zheng Y, Chen D, et al. Chemical composition and skin protective effects of essential oil obtained from ginger (*Zingiber officinale roscoe*). Journal of Essential Oil Bearing Plants. 2018;**21**(6): 1542-1549. DOI: 10.1080/ 0972060X.2018.1533436

[49] Sharma PK, Singh V, Ali M. Chemical composition and antimicrobial activity of fresh rhizome essential oil of *Zingiber officinale* roscoe. Pharmacognosy Journal. 2016;**8**(3):185-190. DOI: 10.5530/ pj.2016.3.3

[50] Ekundayo O, Laasko I, Hiltunen R. Composition of ginger (*Zingiber officinale* Roscoe) volatile oils from Nigeria. Flavour and Fragrance Journal. 1998;**3**(2):85-90. DOI: 10.1002/ ffj.2730030207

[51] Ginger Essential Oil. Available from: https://n-essentials.com.au/ product/ginger-essential-oil/. [Accessed July 18, 2022]

[52] Ginger Root Oil. TGSC Information System. Available from: http://www.thegoodscentscompany.com/data/es1101221.html. [Accessed July 18, 2022]

[53] Kamaliroosta Z, Kamaliroosta L, Elhamirad A. Isolation and identification of ginger essential oil. Journal of Food Bioscience and Technology. 2013;**3**: 73-80. Available from: www.sid.ir/en/journal/ViewPaper.aspx

[54] Hoferl M, Stoilova I, Wanner J, Schimdt E, Jirovetz D, Trifonova D, et al. Compositiion and comprehensive antioxidant activity of ginger (*Zingiber officinale*) essential oil from Ecuador. Natural Products Communications. 2015;**10**(6):1085-1090 PMID: 26197557

[55] Anandaraj M, Devasahayam S, Zachariah TJ, Eapen SJ, Sasikumar B, Thankamani CK. In: Rema J, Madan MS, editors. Ginger (Extension Pamphlet). Kerala, India. 2001. p. 2

[56] Mesomo MC, Corazza ML, Ndaiye PM, Dalla Santa OR, Cardozo L, Scheer AP. Supercritica; CO2 extracts and essential oil of ginger (Zingiber officinale R.): Chemical composition and antibacterial activity. Journal of Supercritical Fluids. 2103;**80**:44-49. DOI: 10.1016/j.supflu.2013.03.031

[57] Stoyanova A, Konakchiev A, Damyanova S, Stoilova I, Suu PT. Composition and antimicrobial activity of ginger essential oil from Veitnam. Journal of Essential Oil Bearing Plants. 2006;**9**(1):93-98. DOI: 10.1080/0972060X.2006.10643478

[58] Kiran CR, Chakka AK, Padmakumari Amma KP, Nirmala Menon A, Sree Kumar MM, Venugopalan VV. Essential oil composition of fresh ginger cultivars from North-East India. Journal of Essential Oil Research. 2013;**25**(5): 380-387. DOI: 10.1080/10412905.2013.796496

[59] Repetto MG, Boveris A. Bioactivity of sesquiterpenes: Compounds that protect from alcohol-induced gastric mucosal lesions and oxidative damage. Mini-Reviews in Medicinal Chemistry. 2010;**10**(7):615-623. DOI: 10.2174/138955710791383992

[60] Santoyo S, Cavero S, Jaime L, Ibañez E, Señoráns FJ, Reglero G. Chemical composition and antimicrobial activity of Rosmarinus officinalis L. Essential Oil Obtained via Supercritical Fluid Extraction Journal of Food Protection. 2005;**68**(4):790-795. DOI: 10.4315/0362-028x-68.4.790

[61] Wohlmuth H, Smith MK, Brooks LO, Myers SP, Leach DN. Essential oil composition of diploid and tetraploid (*Zingiber officinale* Roscoe) grown in Australia. Journal of Agriculture and Food Chemistry. 2006; **54**(4):1414-1419. DOI: 10.1021/jf0521799